APS
Advances in Pharmacological Sciences

Adverse Reactions to Non-Steroidal Anti-Inflammatory Drugs: Clinical Pharmacoepidemiology

Edited by M. Kurowski

Birkhäuser Verlag
Basel · Boston · Berlin

Editor's address
Dr. Michael Kurowski
Institut für klinische Pharmakologie
Charité der Humboldt-Universität Berlin
Schumann-Straße 20/21
D-O-1040 Berlin / FRG

Copy-Editing by Dr. S. Shapiro, Sieberstraße 14, CH–8055 Zürich, Switzerland

Library of Congress Cataloging-in-Publication Data

Adverse reactions to non-steroidal anti-inflammatory drugs : clinical
pharmacoepidemiology / edited by M. Kurowski. – (Advances in pharmacological sciences)
ISBN 3-7643-2628-X (acid-free paper) – ISBN 0-8176-2628-X (acid-free paper)
1. Anti-inflammatory agents – Side effects. 2. Anti-inflammatory
agents – Toxicology.   I. Kurowski, M. (Michael), 1953–   .
II. Series. [DNLM: 1. Anti-Inflammatory Agents, Non-Steroidal – adverse effects.
2. Monitoring, Physiologic.   QV 247 A244]
RM405.A38   1992   615'.7– dc20

Deutsche Bibliothek Cataloging-in-Publication Data

Adverse reactions to non-steroidal anti-inflammatory drugs:
clinical pharmacoepidemiology / ed. by M. Kurowski. – Basel ;
Boston ; Berlin : Birkhäuser, 1992
   (Advances in pharmacological sciences)
   ISBN 3-7643-2628-X
NE: Kurowski, Michael [Hrsg.]

© 1992 Birkhäuser Verlag Basel, P.O. Box 133, 4010 Basel / Switzerland
Printed in Germany on acid-free paper
ISBN 3-7643-2628-X
ISBN 0-8176-2628-X

Table of Contents

Preface

Non-steroidal anti-inflammatory drugs (NSAIDs) are among the most frequently-prescribed drugs in the Western world. During the past decade a large number of new compounds were marketed, some of which had to be withdrawn after a short while because of adverse drug reactions. These experiences led to an understanding that a complete picture of the safety of these drugs can only be obtained *after* their introduction into the marketplace. Different methods of post-marketing surveillance (PMS) serve as important tools for monitoring the frequency of adverse reactions and for generating and corroborating hypotheses.

Experience with PMS has thusfar been quite limited in Germany. In 1987 a non-profit organization, "Verein zur Langzeituntersuchung von Arzneimittelwirkungen auf dem Gebiet der Rheumatologie e.V." (VLAR) was founded by interested physicians and pharmacologists to raise funds and perform investigations on the safety of NSAIDs. The first project by the VLAR, SPALA ("Safety Profile of Antirheumatics in Long-Term Administration"), was sponsored by F. Hoffmann-La Roche AG (Basel, Switzerland). In July 1990, when the project was successfully terminated, almost 30,000 patients had been completely documented and their medical records entered into a computer for subsequent review by a select panel of experts with experience in monitoring adverse reactions to NSAIDs. The comments, criticisms, and ideas of these experts were brought together at a symposium organized by the VLAR at the Klinikum Steglitz (Berlin, Germany) on 12 October 1990. Many persons and institutions contributed to the success of this symposium, particularly Prof. Dr. H. Kewitz (Berlin) who served as the host and chairman, Prof. K. Brune (Erlangen, Germany) who served as a co-chairman, and F. Hoffmann-La Roche AG, which funded the SPALA project. Thanks are also due to the contributors, who presented their experiences with various aspects of PMS and with the clinical recognition of adverse reactions. It is hoped that this symposium will stimulate scientists and manufacturers to continue their efforts in this area of research, which is of such importance to patients and physicians alike.

M. Kurowski, Editor,
for the Executive Board of the "Verein zur Langzeituntersuchung von Arzneimittelwirkungen auf dem Gebiet der Rheumatologie e. V."
February 1991

Adverse Reactions to NSAID: Clinical Pharmacoepidemiology
ed. by M. Kurowski
© 1992 Birkhäuser Verlag Basel

Spontaneous Reporting Systems: Benefits and Pitfalls when Interpreting Safety Data of Non-Steroidal Anti-Inflammatory Drugs

B.-E. Wiholm

Section of Pharmacoepidemiology, Medical Products Agency, P.O. Box 26, S–751 03 Uppsala, Sweden

Introduction

Pain is one of humanity's greatest ailments. Surveys in industrialized countries reveal that on any day 20–30% of the "normal" population suffers from some sort of pain; "rheumatism", joint pains, and muscle pains are the principal complaints. Pain-relieving remedies have a long history, and are one of the most often-used classes of medicines.

The discovery of the pain- and fever-relieving effects of the bark of the salix bush *Salix purpurea* was one of the major contributions to the alleviation of human suffering. However, adverse effects of salicylates have been documented since at least 1798 [1]. Hemorrhagic erosions in the stomach after intake of salicylic acid were also described long ago [2], but considering the doses used at the time (up to 1 g hourly), one should probably regard those early therapeutic trials as clinical toxicology tests rather than pharmacotherapy. It is sobering to reflect upon the fact that also in the 1980s several non-steroidal anti-inflammatory drugs (NSAIDs), which had passed modern regulatory evaluation programs, were withdrawn from the market after a short period because of adverse effects. In the case of benoxaprofen, the doses recommended for a major target group, the elderly, were far too high, as it was not recognized that elderly patients have a significantly decreased capacity to eliminate this drug. When indomethacin was presented in a novel formulation based on a non-dissolving capsule with a laser hole through which the active ingredient together with potassium was slowly released, a sudden and unforeseen outbreak of gastrointestinal (GI) perforations occurred, causing the drug's rapid withdrawal from the market. At least 17 out of a 100 new NSAIDs developed during the last 20 years have been withdrawn because of problems with

2

adverse reactions [3]. In some cases it has been claimed that the withdrawals were based on a small number of spectacular adverse drug reactions (ADRs) instead of a proper estimate of the incidence of ADRs [4]. This points to a general and severe problem in the evaluation of adverse drug effects: a lack of adequate quantitative information.

During clinical trials, only reactions occurring in at least 0.5% of the patients can usually be detected and quantified because of inherent limitations in the size of trials. Moreover, most studies are short, and the number of patients followed for more than six to twelve weeks is often limited. Lastly, the patients in these studies often are not representative of the population taking the drug in clinical practice because of restrictions in age range and the occurrence of concomitant diseases.

Most serious ADRs are presented as single case reports or as studies on a small number of patients, and the number of properly-conducted large epidemiological studies is limited. Until quite recently, safety evaluations were often made on the basis of information from spontaneous reporting systems. This article is concerned with an analysis of some benefits and pitfalls of using spontaneous reporting systems to derive safety data for NSAIDs. The analysis is based on the Swedish reporting system, and the validity of the results will be discussed and compared to data from the literature.

Materials and Methods

The Swedish ADR reporting system

In Sweden a system for monitoring spontaneous reports of ADRs was set up in 1965. At first reporting was entirely voluntary, but since 1975 doctors and dentists have been obliged to report suspected severe or fatal drug reactions as well as drug reactions thought to be new and/or unexpected. A preliminary evaluation of these reports is made by the physicians of the ADR section. For most serious and all fatal cases full medical records, including laboratory tests and autopsy reports, are requested. The information so collected is discussed biweekly at working sessions and quarterly by the full Adverse Reactions Advisory Committee (SADRAC). No special algorithm is routinely used for causality assessment, but the following points are considered for each report:

1. Is there a reasonable temporal connection between drug intake and the suspected reaction?
2. Is there a reasonable pharmacological explanation for the reaction, or has the reaction been described before?

3. Does the reaction diminish or disappear when the dose of the drug is reduced or suspended?
4. Can the patient's primary disease elicit similar symptoms?
5. Can some other drug taken by the patient cause the same symptom, or can the symptom arise from a drug interaction?
6. Did the same symptom occur on previous exposures to the drug, or did it recur upon rechallenge with the drug?

Only reactions believed to have a probable or possible causal connection to the drug are further considered.

Drug sales and prescription registers

Total drug sales statistics are available and have been computerized since 1972. These statistics show the total amount of every drug sold in each pharmacy in terms of either volume, monetary value, or "defined daily doses" (DDDs). The DDD is the estimated average daily dose of a drug for adults when that drug is used for its main indication [5]. The number of DDDs sold per 1,000 inhabitants per day is a gross but useful measure of drug consumption by the public. For comparison with ADRs, the total number of DDDs (in millions) per annum is an appropriate measure. The DDDs for the drugs considered in this article are shown in Table 1.

Table 1. Non-steroidal anti-inflammatory drugs available in Sweden.

Generic name	**Year of first license**	**DDD (mg)**
Phenylbutazone	Feb 1956	300
Oxyphenbutazone*	Sep 1963	300
Indomethacin	May 1965	100
Ibuprofen	Mar 1975	1,200**
Naproxen	Mar 1975	500
Azapropazone	Jan 1978	750
Diclofenac	Dec 1981	100
Piroxicam	Dec 1981	20
Sulindac	Oct 1982	400
Ketoprofen	Mar 1983	150

* Oxyphenbutazone was withdrawn from the market by the manufacturer in July 1984.
** Until 1985 the DDD was 800 mg.

4

During that period, most pharmacies had manual routines. The prescriptions must be stored for 3 years and are ordered sequentially. Thus one of the pharmacists picked out every 288th prescription from the sequentially ordered files. From 1980 onwards the pharmacies have been computerized and the information on the prescription is entered into the computer which now automatically stores the data from every 22nd prescription in a separate file which is transmitted to HQ regularly.

Between 1974 and 1982 most pharmacies used manual record-keeping routines. Prescriptions had to be ordered sequentially and stored for three years; pharmacies selected every 288th prescription from the sequentially-ordered files, and information about age and sex of the patient and the name, amount, and daily dosage of the drug was coded and computerized. From this survey the average or median prescribed daily dose (PDD) could be calculated by age and sex. Since 1980 there has been a trend towards the computerization of pharmacy prescription records. Beginning in 1983 the sampling frequency was increased to 1 in 25 by introducing into pharmacies a computer system which automatically transfers into a separate file information about every 25th prescription filled, the information then being regularly transmitted to the relevant regulatory authorities. However, information on the PDD was lost in the computerization process. The number of prescriptions per 1,000 inhabitants and per year is used as an estimate of the prescribing pressure on the population.

Since 1978 the Diagnosis and Therapy Survey has been a collaborative effort between the drug industry, the National Corporation of Swedish Pharmacies, the Swedish Medical Association, and the National Board of Health and Welfare. In this survey a random sample of physicians each week register all prescriptions on special forms which also state the indication for each drug treatment. Data from these registers are published annually [6].

Adverse reactions

Adverse reactions can be classified according to several different principles, e.g. the organ/system affected or the proposed mechanism. There have been numerous efforts to invent sophisticated mechanism-based ADR classifications, but they have had limitations and errors arising from our lack of knowledge. In a simple and practical way ADRs can be divided into two types:

A. Type A (augmented) reactions are usually directly related to the pharmacody-namic actions of the drug through effects on receptors (A1), or to chemical properties of the drugs (A2). An example of a type A ADR is esophageal ulcerations caused by NSAIDs. Most type A reactions are clearly related to the dose or concentration of the causal drug.

B. Type B (bizarre) reactions have no obvious connection to the pharmacody-
    namic effects of the drug, nor have they any clear and simple relationship to
    dosage. Some type B ADRs are of immunologic origin, whereas others are
    caused by the formation of aberrant toxic metabolites. For many type B ADRs
    the underlying mechanism is not understood at all.

NSAIDs

Almost all NSAIDs are weak organic acids, or salts and esters of weak organic
acids. They fall into 7 different groups, 6 of which are marketed in Sweden
(Fig. 1). However, NSAIDs all share at least one important pharmacologic effect,
viz. the inhibition of cyclooxygenase, which diminishes the production of prostaglandins. Thus, all NSAIDs share some type A adverse effects while differing in
their propensities to produce type B adverse reactions. Table 2 lists some examples
of important ADRs reported for NSAIDs.

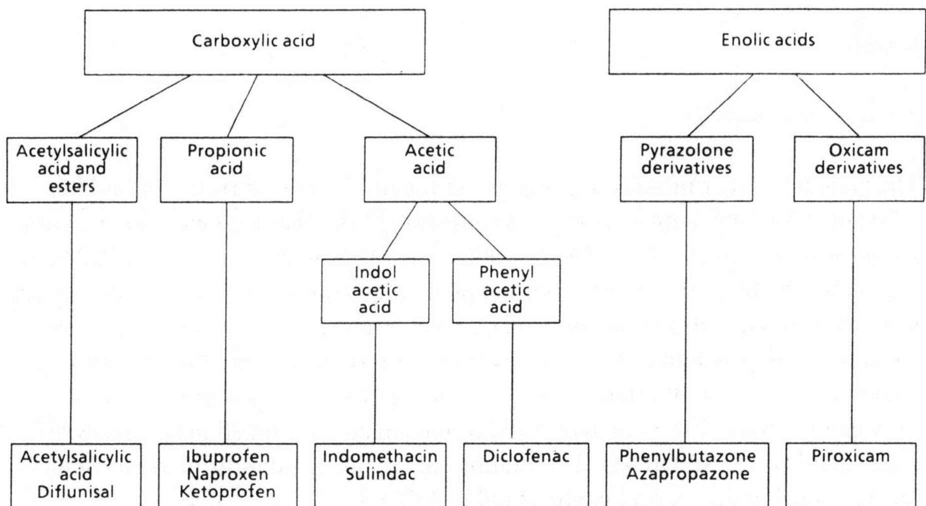

Fig. 1. Schematic classification of non-steroidal anti-inflammatory drugs available in Sweden.

6

Table 2. Important adverse effects of NSAIDs.

Organ system	Type	Symptom/Reaction
General	B	Fever, lymphadenopathy, general hypersensitivity
Blood	B	Agranulocytosis, thrombocytopenia, aplastic anemia
	A	Thrombasthenia (increased bleeding time resulting from decreased thrombocyte function)
Circulation/Kidney	A	Salt and water retention – cardiac failure
	A	Reversal of antihypertensive effect of diuretics
	A	Renal papillary necrosis
	A + B	Nephritis, nephrosis
CNS	A or B	Headache, vertigo, confusion, hallucinations, anxiety, depression, amnesia
	B	Aseptic meningitis
Upper GI	A	Erosive gastritis, ulcers and bleeding
Lower GI	A	Ulcers and perforation, precipitation of colitis?
Liver	B	Hepatitis and cholestatic reactions
Lung	B →A	Bronchoconstriction
Skin	B	Rash, urticaria, mucocutaneous syndrome, toxic epidermal necrolysis
Others	B	Sialadenitis, pancreatitis, opticus neuritis

Results

Number of reports

The annual number of ADR reports has gradually increased from 160 in 1965 to 3,000 in 1988. The number of reports concerning NSAIDs has increased concomitantly with the introduction of new products into the market (Fig. 2). In 1982 there was a remarkable jump of 300 over the preceding year's number of ADR reports when diclofenac, piroxicam and sulindac were introduced. Most of these reports concerned GI problems and skin reactions to piroxicam. In 1982 NSAIDs accounted for 15% of all ADR reports, in contrast to a previous average of 4–7%. Between 1975 and 1989 a total of 2,237 reports of adverse reactions to *all* NSAIDs were received by SADRAC. The number of reports on the different substances received during this period is presented in Table 3.

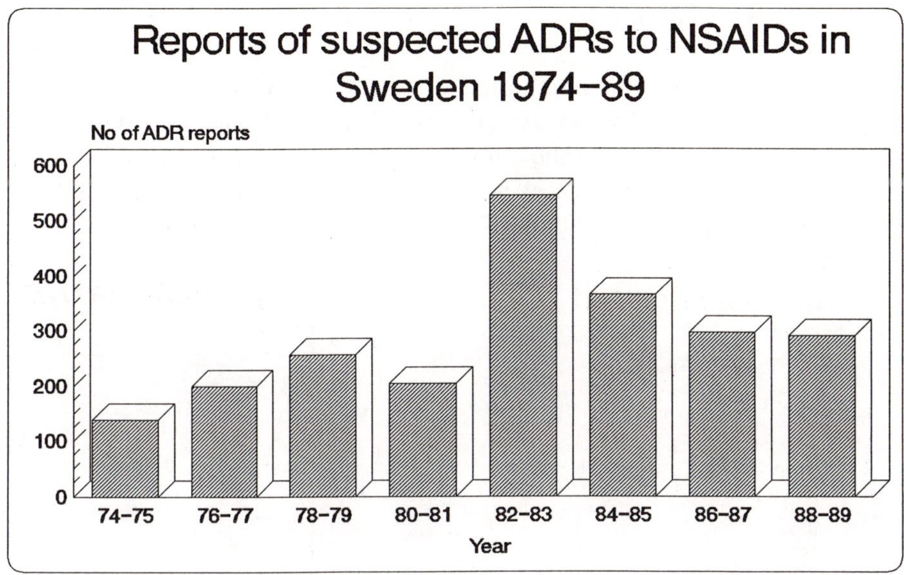

Fig. 2. Number of ADR reports for non-steroidal anti-inflammatory drugs in Sweden, 1974–1989. Each bar represents the mean of two years.

Table 3. Number of reports and adverse reactions for NSAIDs[*] in Sweden, 1975–1989.

NSAID	Time period	Number of reports	Number of reactions	Fatal cases
Phenylbutazone	1975–1989	89	125	3
Oxyphenbutazone	1975–1984	185	277	3
Indomethacin	1975–1989	320	403	13
Naproxen	1975–1989	414	547	9
Ibuprofen	1975–1989	273	340	5
Azapropazone	1978–1989	103	130	1
Diclofenac	1982–1989	214	285	8
Piroxicam	1982–1989	357	430	2
Sulindac	1982–1989	257	394	3
Total		2,212	2,931	47

*Excluding ketoprofen and acetylsalicylic acid

Profiles of adverse reactions

Qualitatively, the ADR patterns reported for different NSAIDs have many features in common vis-à-vis type A reactions like GI ulcers and bronchoconstriction and presumed type B reactions like those involving the blood, liver and skin. However, there are some distinct differences in the profiles of reported reactions between different NSAIDs (Fig. 3). Some of these differences appear to be real, e.g.

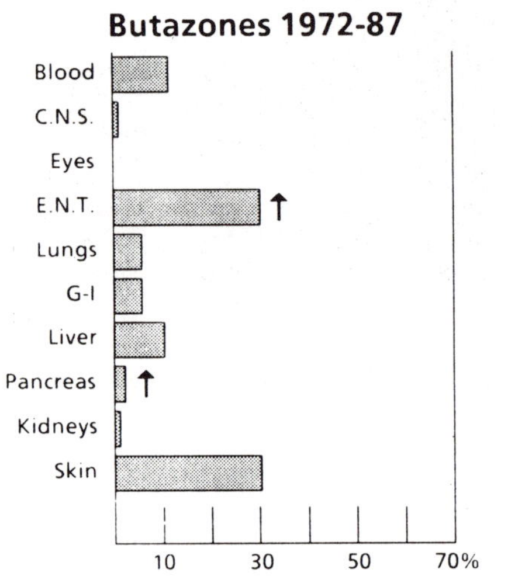

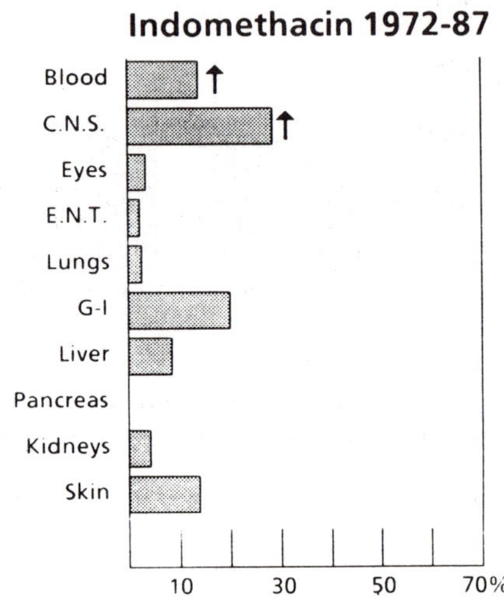

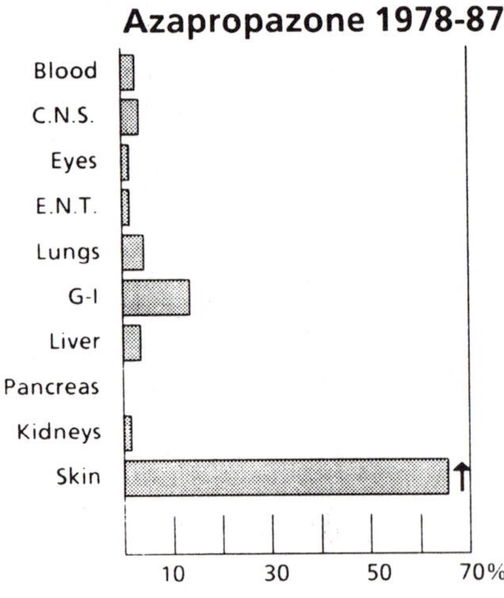

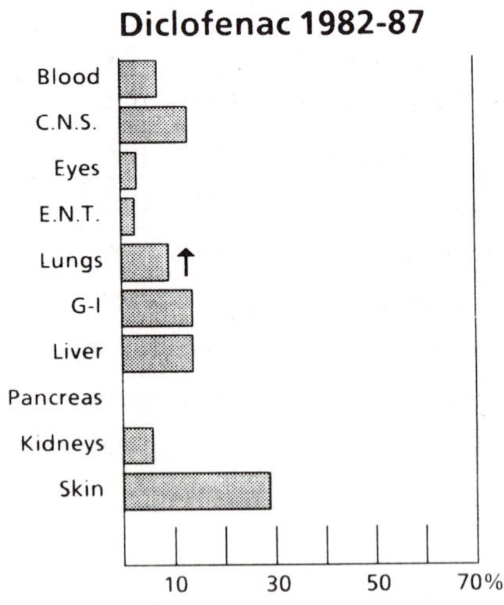

Fig. 3 a)

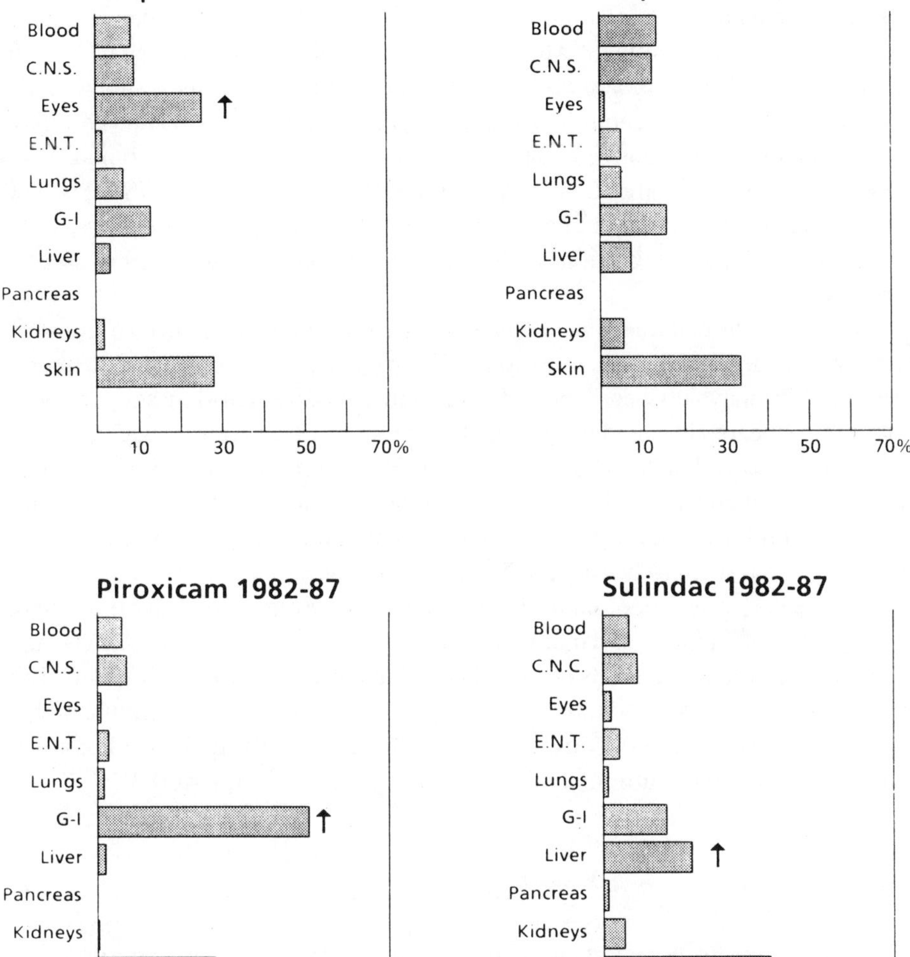

Fig. 3. a) and b) Profiles of adverse reactions reported for non-steroidal anti-inflammatory drugs in Sweden. Arrow denotes the drug with the largest proportion within the system/organ class.

patterns that have also been found in other spontaneous reporting systems but which did not arouse much interest in the medical or lay press. Some other differences may be due primarily to reporting habits.

10

The highest proportion of blood dyscrasias is reported for indomethacin, the butazones, and naproxen. For indomethacin and the butazones the relative risk of blood dyscrasias is prominent when viewed in relation to sales data as well as in the context of formal epidemiologic studies [7]. Central nervous system (CNS) reactions are also conspicuous in the reporting profile for indomethacin. This may be due in part to the negative influence of indomethacin on cerebral blood flow. Confusion in elderly patients dominates the CNS reactions to all NSAIDs, but there are some reports of nightmares, amnesia, hallucinations, and even psychotic reactions, as well as convulsions and (at least for ibuprofen) rare cases of aseptic meningitis.

Among ocular reactions ibuprofen is featured prominently. However, most of these reports describe vague sensations of blurred vision, diplopia, and disturbed color perception which were reported after claims of optic neuritis associated with ibuprofen appeared in a medical article. These reports may therefore be a manifestation of the so-called "band wagon" effect. The ear, nose, and throat reactions reported do not consist of the typical dose-related tinnitus or deafness which are well known for aspirin; rather, they consist mostly of cases with sialadenitis and parotitis, which seem to be true ADRs of the butazones.

Liver reactions are reported for all NSAIDs, but they are especially prominent in the profile of sulindac and diclofenac. Sulindac is also the only NSAID beside the butazones for which cases of pancreatitis have been reported in Sweden. The high proportion of skin reactions for azapropazone is probably a true feature of this pyrazolone, as it has been shown to accumulate in the skin and to undergo phototoxic reactions in a high proportion of takers under controlled conditions [8,9].

ADRs in relation to sales and prescriptions

The incidence of ADR reports in relation to sales varies markedly over time and from drug to drug (Fig. 4). The level of reporting is generally higher for newer drugs, and there is a marked peak in reporting during the first two to five years after initial marketing. An exception to this pattern is the butazones, which were introduced more than 30 years ago but recently have shown an increased reporting rate. This may be due mostly to the intense debate about several serious adverse effects of these drugs both in the medical and lay press. Such phenomena make overall comparisons between different drugs difficult to interpret. The average incidence of all reports between 1975 and 1989 varies from 1.4 to 15.2 per million DDDs (Tab. 4), corresponding to 5–54 reports per 10,000 patient-years if the DDD values are correct. When compared to the situation in clinical trials, where from 2% to 10% of the patients drop out because of ADRs, the enormous degree of underreporting of minor reactions becomes obvious.

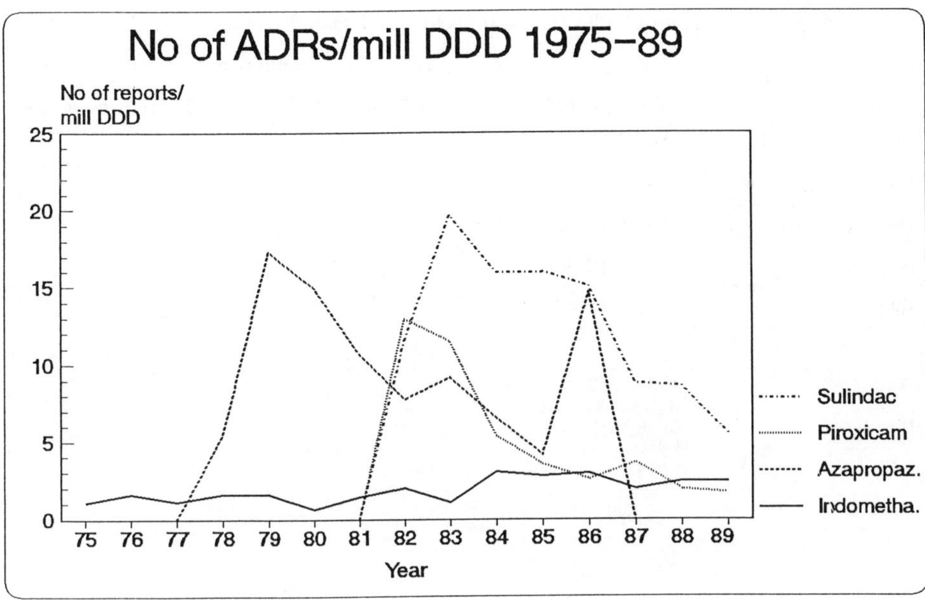

Fig. 4 a)

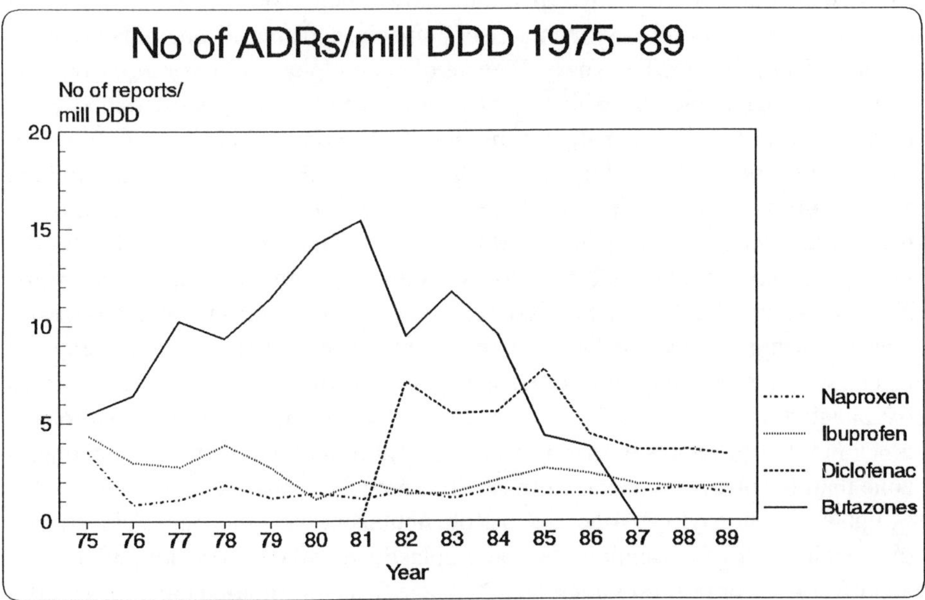

Fig. 4. a) and b). Incidence of adverse reactions in relation to sales of non-steroidal anti-inflammatory drugs in Sweden.

Table 4. Reported incidence of adverse reactions to NSAIDs in Sweden.

NSAID	Time period	Total No. of reports per million DDDs	10,000 treatment yrs
Phenylbutazone	1975–1989	5	18
Oxyphenbutazone	1975–1984	15	55
Indomethacin	1975–1989	2	6
Naproxen	1975–1989	1	5
Ibuprofen	1975–1989	2	8
Azapropazone	1978–1989	12	44
Diclofenac	1982–1989	5	17
Piroxicam	1982–1989	9	33
Sulindac	1982–1989	13	47

Incidence of adverse reactions for some major organ systems

A. Blood and bone marrow

In spontaneous reporting systems blood dyscrasias such as leucopenia and agranulocytosis, thrombocytopenia, pancytopenia, and aplastic anemia have been reported for most NSAIDs. The incidences of agranulocytosis, thrombocytopenia, and pancytopenia (including bicytopenia and aplastic anemia) reported in Sweden in relation to sales data are shown in Table 5. The denominator used, "100,000 treatment years", is derived from the total number of PDDs sold per year divided by 365. From these data it is seen that blood dyscrasias related to NSAIDs are rare. At face value the risk per 100,000 treatment years seems to be higher for the metabolite oxyphenbutazone than for the substrate compound phenylbutazone. This can be explained by the fact that, if it is to occur at all, agranulocytosis usually appears during the first weeks of treatment, and only rarely thereafter. Phenylbutazone was usually used for long-term treatment of rheumatoid arthritis, whereas oxyphenbutazone was principally used for short periods of time. Therefore, a given number of DDDs would translate into more individuals for oxyphenbutazone than for phenylbutazone. If these facts are taken into account, then the risk estimates for phenylbutazone and oxyphenbutazone become much closer. The total incidences of agranulocytosis and aplastic anemia among out-patients are about 7 and 2 per million inhabitants per year. This information can also be used to calculate rough values for relative and excess risk from spontaneous reports and sales data. The estimates of risk for butazone-induced agranulocytosis from

Table 5. Reported incidences of blood dyscrasias in Sweden.

NSAID	Time period	Reports No.	Incidence: no. of reports per 100,000 treatment years			
			Agranulo-cytosis	Thrombo-cytopenia	Pancyto-penia	Total
Oxyphenbutazone	1975–1984	12	9	21	6	36
Diclofenac	1982–1989	14	3	6	2	11
Phenylbutazone	1975–1989	10	2	6	12	20
Piroxicam	1982–1989	10	3	6	2	11
Sulindac	1982–1989	7	6	6	2	14
Indomethacin	1975–1989	36	3	3	1	7
Naproxen	1975–1989	35	1	3	1	5
Azapropazone	1978–1989	1	–	–	4	4
Ibuprofen	1975–1989	9	1	1	1	3

the spontaneous reporting system and those derived by the International Agranulocytosis and Aplastic Anemia study (IAAAS) [7] are rather similar (Tab. 6). In the IAAAS an elevated relative risk of agranulocytosis was also identified for indomethacin.

For aplastic anemia, an increased risk was identified for the butazones (relative risk, 9), indomethacin (RR, 13) and diclofenac (RR, 9) but the attributable risk was very low. More data will emerge from the IAAAS, which is continuing, and elevated risks for other NSAIDs will probably emerge as well.

Table 6. Comparison of risk estimates for sulfa-induced agranulocytosis from the Swedish drug monitoring system and the International Agranulocytosis and Aplastic Anemia Study. Relative risk (95% confidence limits).

Source	Butazones	Indomethacin	Sulfasalazine	Trimethoprim-sulfametoxazole	Thyrostatics*
SDMS	5.5 (1.4–17.6)	3.0 (1.7–5.3)	107 (67–170)	17 (8–37)	163 (135–197)
IAAAS	3.9 (1.4–11.0)	6.6 (2.6–17.0)	123** (16–966)	12 (4–40)	97 (36–262)

* propylthiouracil, carbimazole, tiamazole.   ** Swedish part of IAAAS.

14

B. GI ulcers and bleeding

Dyspepsia, heartburn, and GI upset are probably the most common adverse reactions to aspirin and other NSAIDs. From gastroscopic studies it is known that a majority of patients taking these preparations will initially show mucosal irritation, microbleeding, and small superficial ulcers [10]. In most cases, however, these gastroscopic findings decrease during continued treatment. The relation between these findings and the development of bleeding and perforated ulcers is unclear.

An abundance of studies on the relationship between NSAIDs and the risk of upper GI ulcers and bleeding have appeared in the medical literature. The results of most of these studies are difficult to interpret in a meaningful way. Studies based on spontaneous reports on this subject probably have more severe reporting biases than in most other symptom areas. Many formal studies suffer from inadequate, imprecise, or varying definitions and ascertainment of exposure. Control selection is another major problem in several studies, and it becomes very difficult to interpret results from studies where cases with prior ulcer disease are analyzed together with those experiencing their first episode of ulcer or bleeding. The biased spontaneous reporting is shown in Table 7, where the risk seems to be much higher for piroxicam than for other NSAIDs. Moreover, in epidemiological studies aspirin has a higher risk than many modern NSAIDs, though less than ten aspirin ulcers have been reported since 1965.

Table 7. Reported incidence of GI ulcers and bleeding in Sweden.

| Generic name | Total number | Incidence per 100,000 treatment years | | Bleedings unspecified | Total |
| | | Ulcers | | | |
		Uncomplicated	Bleeding and perforations		
Piroxicam	118	75	47	28	151
Azapropazone	7	5	28	9	42
Sulindac	5	3	9	3	15
Phenylbutazone	11	–	1	10	11
Oxyphenbutazone	7	1	1	6	9
Indomethacin	22	1	2	2	5
Diclofenac	1	–	3	–	3
Ibuprofen	3	–	0.7	0.3	1

This situation was probably largely created by the marketing division at Pfizer. Piroxicam was extremely heavily marketed as a new "safe" NSAID which could be taken once a day without side effects, which in turn led to overprescribing. The first patients to be prescribed piroxicam were probably those who previously did not tolerate the older preparations; of course, some of these patients did not tolerate piroxicam either, and they developed ulcers and GI bleeding. When surgeons who cared for the bleeding patients realized that piroxicam was just another NSAID with the same ulcer risk as other "modern" NSAIDs, they felt deceived by the marketing people and filed ADR reports. In fact, there is some evidence that piroxicam has a *greater* tendency to induce ulcers than some other NSAIDs. Piroxicam has a very long half-life, and if its ulcerogenic tendencies are linked to the inhibition of prostacyclin (prostaglandin I_2) then any drug which continuously inhibits prostacyclin formation should also exhibit a higher GI risk than a drug with a short half-life. In all published data based on spontaneous reporting systems, there is a strong correlation between the "incidence" of GI adverse effects and the serum half-life of NSAIDs [11–14].

C. Liver reactions

Liver reactions have been reported for all NSAIDs available in Sweden (Tab. 8). Elevated aminotransferase levels are common when aspirin is used in high doses for the treatment of rheumatoid arthritis, especially in children. Low serum albumin levels also seem to increase the risk for this reaction. Apart from benoxaprofen, liver reactions seem to be rare for the newer generation of NSAIDs, though the only published studies of their frequencies were based on spontaneous

Table 8. Reported incidence of liver reactions in Sweden.

NSAID	Time period	Number of reports	No. of reports per 100,000 treatment yrs
Sulindac	1982–1989	71	131
Oxyphenbutazone	1975–1984	21	63
Diclofenac	1982–1989	35	30
Phenylbutazone	1975–1989	14	28
Azapropazone	1978–1989	4	17
Piroxicam	1982–1989	8	7
Indomethacin	1975–1989	24	5
Naproxen	1975–1989	30	4
Ibuprofen	1975–1989	13	4

reports as per the Swedish example. There is a recent report from France describing 50 cases of hepatitis due to NSAIDs in 1985 [15]. The estimated frequency of these reactions varied between 1 in 50,000 patients and 1 in 500,000 patients. Swedish reports based on the DDD data, indicate that liver reactions occur in from a few to about 150 cases per 100,000 treatment-years. Whether this frequency range reflects true differences between the various NSAIDs is, however, a subject requiring further study.

D. Renal reactions

The pharmacological effects of NSAIDs can lead to fluid retention and decreased glomerular blood flow, especially in patients with cardiac or renal failure. A number of other renal effects have also been described in single case reports, but the incidence of renal toxicity associated with NSAIDs is not well known. In one of the few published epidemiologic studies on this problem Guess et al. [16] compared hospitalization rates for various renal diseases for 134,000 NSAID users and 848,000 nonusers in the province of Saskatchewan, Canada. Significant risks were identified only for hyperkalemia (relative risk, 19; excess risk, 2 per 10,000 treatment-years) and nephritis and nephropathy (World Health Organization International Classification of Diseases, 9th edition, no. 583) (relative risk, 7.6; excess risk, 3.2 per 10,000 treatment-years). The relative risk for acute renal failure was insignificantly increased in NSAID users. Even in this large study it was not possible to examine whether different NSAIDs carried different risks for renal reactions.

Table 9. Reported incidence of Stevens-Johnson Syndrome and toxic epidermal necrolysis in Sweden.

NSAID	Time period	Number of reports	No. of reports per 100,000 treatment yrs
Oxyphenbutazone	1975–1984	15	45
Sulindac	1982–1989	15	28
Phenylbutazone	1975–1989	3	6
Piroxicam	1982–1989	5	5
Naproxen	1975–1989	8	1
Ibuprofen	1975–1989	6	2
Diclofenac	1982–1989	3	2
Indomethacin	1975–1989	4	1
Azapropazone	1978–1989	–	–

E. Serious skin reactions

Stevens-Johnson syndrome and toxic epidermal necrolysis are very rare but serious skin diseases which in Sweden occur at a frequency of ca. 2–4 cases per million inhabitants per year. Several drugs have been implicated in eliciting such reactions, notably long-acting sulfonamides, anti-epilepsy drugs, and NSAIDs. The number of cases reported in relation to sales is depicted in Table 9. As with liver reactions, oxyphenbutazone and sulindac top the list, but in the absence of formal studies one must be extremely cautious in the interpretation of these data.

Discussion

A spontaneous reporting system can be regarded as an incomplete (and, at worst, biased) case series without any information on the size or characteristics of the population exposed to the drug except for that inferred by the indication for treatment. Under these circumstances it is hardly possible to establish a causal connection between an adverse event and a drug unless:

(*a*) there is at least one case with positive rechallenge and some more supportive cases without known confounding drugs or diseases; or

(*b*) there is a cluster of exposed cases reported where the background incidence of the adverse event is close to zero, and there is no confounding.

Even the reappearance of an adverse event upon readministration of the drug is no proof of causality [17], though in practice one can make a strong case for a causal connection if there is a cluster of cases with good clinical information and in which the same event has reappeared at least once in each patient, and if the event diminishes or disappears after withdrawal of the drug and does not spontaneously reappear. It is also clear that unless the total incidence of ADRs is equal for, e.g., ibuprofen and the butazones, the different proportions of events such as blood dyscrasias in ADR profiles cannot be interpreted as differences in risk. For such comparisons to be valid, estimates of the denominators (numbers of users) are needed.

Many drug regulatory authorities and pharmaceutical manufacturers have access to information from which they can project both the size and characteristics of the exposed population and the background incidences of diseases. If the rate of reporting is known, the estimate of the numerator (number of cases) becomes more accurate. From studies using registers of hospital discharge diagnoses it has been possible to calculate reporting rates for some geographical areas, ADRs, and periods of time. Between 20% and 40% of serious reactions (blood dyscrasias, thromboembolic disease, Stevens-Johnson syndrome, etc. [18]), identified by

18

checking medical records of patients discharged with these diagnoses, are usually reported to SADRAC [14]. As a case in point, by checking all positive *bacille Calmette-Guérin* cultures in bacteriology laboratories it was found that almost 80% of all children who developed an osteitis after BCG vaccination had been reported [19].

However, reporting rates cannot be generalized. They are important to know when evaluating data but should not be used to correct calculations for under-reporting, because the data may well be drug-, time-, area-, and/or reaction-specific. By knowing the number of DDDs sold and the average PDD it is possible to roughly estimate the total person-time of exposure for a particular drug.

If prescription statistics are available, the number of cases reported per prescription may actually be a better estimate of the risk to outpatients than if the number of treatment weeks are calculated from sales data, at least for antibiotics where doses and treatment times may vary with patient age and indication.

If information from an efficient spontaneous reporting system can be combined with drug sales and prescription statistics, it may be possible to derive rough estimates of the frequency or incidence rate of an ADR. Such estimates will not be as accurate as those derived from clinical trials or formal epidemiological post-marketing surveys; however, they can serve as a first indicator of the size of a potential problem and for very rare reactions they may be the *only* possible measure.

Spontaneous reporting is recognized as the most effective method to discover rare but serious adverse reactions, but it is not thought to yield valid estimates of frequency or risk. This is probably most often the case when considering drug-related beneficial or adverse effects in situations where the ADR is a type A reaction, as type A reactions are the most common. There are situations, however, where the ADR of concern is a rare type B reaction, for which the background incidence of the disease is low, and where it seems possible to obtain valid data with the approach outlined in the examples above.

If the background incidence of a disease is known or can be estimated from other sources, it is sometimes possible to obtain rough estimates of rate ratios and rate differences from spontaneously-reported data on ADRs and sales and prescription statistics. This technique was first applied in 1983 during an investigation of a possible relationship between a new anti-depressant drug and the development of Guillan-Barré syndrome in patients experiencing flulike hypersensitivity reactions from this new medicine [20].

Risk estimates derived from the Swedish spontaneous monitoring system and from an international case-control study of blood dyscrasias were compared (Tab. 6). In both cases the estimates of relative and excess risks were similar enough to have led to the same clinical and regulatory evaluations of butazones, cotrimoxazole, and sulfasalazine.

Conclusion

The number of ADRs reported annually for NSAIDs has increased with time, and the total number is now quite substantial. To evaluate the implications of these reports it is imperative to know the size and characteristics of the population exposed to the drugs as well as the rate of reporting. As a consequence of underreporting and selective reporting, results based on studies of spontaneous reports must always be interpreted cautiously and should be supplemented by formal epidemiologic studies. Unfortunately, for many serious ADRs reported for NSAIDs, formal studies are lacking. Generally speaking, serious adverse reactions to NSAIDs are rare, and the NSAIDs available in Sweden are acceptably safe if used with proper care.

References

1. Longmore, G., Account of fourteen men of the Royal artillery at Quebec, who were nearly poisoned by drinking a decoction of certain plants. Ann. Med. Edinburgh **3**: 364 (1798).
2. Myers, A. B. R., Salicin in acute rheumatism. Lancet **2**: 364 (1876).
3. Rainsford, K. D., Introduction and historical aspects of the side-effects of anti-inflammatory analgesic drugs. In: Rainsford. K. D. and Velo, G. P., Eds., Side-effects of anti-inflammatory drugs, Part I. MTP Press, Lancaster 1987, pp. 3–26.
4. Inman, W. H. W., Risks in medical intervention. Prescript. Event Monitor. News **2**: 16–36 (1984).
5. Nordic Council on Medicines, Nordic statistics on medicines 1978–1980, Part II. Nordic Council on Medicines, Uppsala 1982.
6. National Corporation of Swedish Pharmacies, Swedish drug statistics 1989 (Svensk läkemedelsstatistik). National Corporation of Swedish Pharmacies, Stockholm 1989.
7. The International Agranulocytosis and Aplastic Anemia Study, Risks of agranulocytosis and aplastic anemia. A first report of their relation to drug use with special reference to analgesics. J. Am. Med. Assn. **256**: 1749–1757 (1986).
8. Olsson, S., Biriell, C., and Boman, G., Photosensitivity during treatment with azapropazone. Br. Med. J. **291**: 239 (1985).
9. Jones, R.A., Navaratnam, S., Parsons, B. J., and Philips, G. 0., Photosensitivity due to anti-inflammatory analgesic drugs: a laser flash photolysis study of azapropazone. In: Rainsford, K. D. and Velo, G. P., Side-effects of anti-inflammatory drugs, Part II. MTP Press, Lancaster 1987, pp. 345–354.
10. Caruso, I. and Porro, G. B., Gastroscopic evaluation of antiinflammatory agents. Br. Med. J. **280**: 75–78 (1980).
11. Kromann-Andersen, H., Kovacs, I., and Pedersen, A., Biverkningar under brug af ikke-steroide antireumatika i Danmark igennem 15 ar (Adverse reactions during non-steroidal anti-rheumatics in Denmark during a 15-year period). Ugeskr. Laeqer. **148**: 462–468 (1986).
12. CSM update, Non-steroidal anti-inflammatory drugs and serious gastrointestinal adverse reactions–2. Br. Med. J. **292**: 1190–1191 (1986).
13. Paulus, H. E., FDA Arthritis Advisory Committee meeting: post-marketing surveillance of nonsteroidal antiinflammatory drugs. Arthritis Rheum. **28**: 1168–1169 (1985).

14. Wiholm, B.-E., Myrhed, M., and Ekman, E. (1987), Trends and patterns in adverse drug reactions to non-steroidal anti-inflammatory drugs reported in Sweden. In: Rainsford. K. D. and Velo, G. P., Side-effects of anti-inflammatory drugs, Part I. MTP Press, Lancaster 1987, pp. 55–72.

15. Castot, A., Netter, P., Larrey, D., Carlier, P., Gaira, M. et al., Hepatitis due to non-steroidal anti-inflammatory drugs–a cooperative study by the regional centres of drug surveillance for the year 1985. Therapie **43**: 229–233 (1988).

16. Guess, H. A., West, R., Strand, L. M., Helston, D., Lydick, E., Bergman, U., and Wolski, K., Hospitalizations for renal impairment among users and non-users of non-steroidal anti-inflammatory drugs in Saskatchewan, Canada 1983. In: Rainsford, K. D. and Velo, G. P., Side-effects of anti-inflammatory drugs, Part II. MTP Press, Lancaster 1987, pp. 367–375.

17. Rothman, K. J., *Modern epidemiology*. Little, Brown, Boston 1986.

18. Wiholm, B.-E., Spontaneous reporting of ADR. In: Werkü, L. (ed.), Detection and prevention of adverse drug reactions (Skandia International Symposia). Almqvist & Wiksell, Stockholm 1983, pp. 152–167.

19. Bottiger, M., Romanus, V., de Verdier, C., and Boman, G., Osteitis and other complications caused by generalized BCG-itis. Experiences in Sweden. Acta. Pediat. Scand. **71**: 471–478 (1982).

20. Fagius, J., Osterman, P. 0., Siden, A., and Wiholm, B.-E., Guillain-Barre syndrome following zimeldine treatment. J. Neurol. Neurosurg. Psychiatry **48**, 65–69 (1985).

Adverse Reactions to NSAID: Clinical Pharmacoepidemiology
ed. by M. Kurowski

Drug Utilization Studies: A Useful Tool for the Safety Evaluation of Non-Steroidal Anti-Inflammatory Drugs?

P. K. M. Lunde

Department of Pharmacotherapeutics, University of Oslo, P.O. Box 1065 Blindern, N–0316 Oslo 3, Norway

Introduction

A complex terminology is presently used to indicate various approaches to the post-registration monitoring of drug exposure, according to extent, circumstances, and subsequent consequences of such exposure. Terms currently in vogue include drug utilization studies (DUS), pharmacoepidemiology, post-marketing surveillance (PMS), phase IV studies, and health service research. Although these terms are somewhat differently defined [1,2] and often inconsistently perceived, the concepts and objectives are more or less similar, viz. to contribute to our understanding of benefit/risk and cost/effectiveness ratios of drug interventions and the interrelationship of these ratios (Fig. 1). In relevant contexts the scope of post-marketing monitoring may be expanded to include non-drug interventions of a diagnostic, preventive, curative, and/or symptomatic nature.

The comprehensive definition of drug utilization [3], i.e. "the marketing, distribution, prescription and use of drugs in a society, with special emphasis on the resulting medical, social and economic consequences" calls for a "complete" medico-social and health economic audit [4–6] involving:

1. collection and compilation of all relevant facts and premises;
2. data organization and analysis by competent auditors; and
3. decisions at various levels of the drug and health care chain, implementation of these decisions, and their perireevaluation and (if necessary) adjustment.

In other words, the general aims of drug utilization studies are (*a*) to identify and define the problem(s); (*b*) to analyse problems according to their significance, causality, and consequences; (*c*) to establish a weighted basis for decisions and solutions to problems (Fig. 1); and (*d*) to assess the effects of the actions and interventions taken.

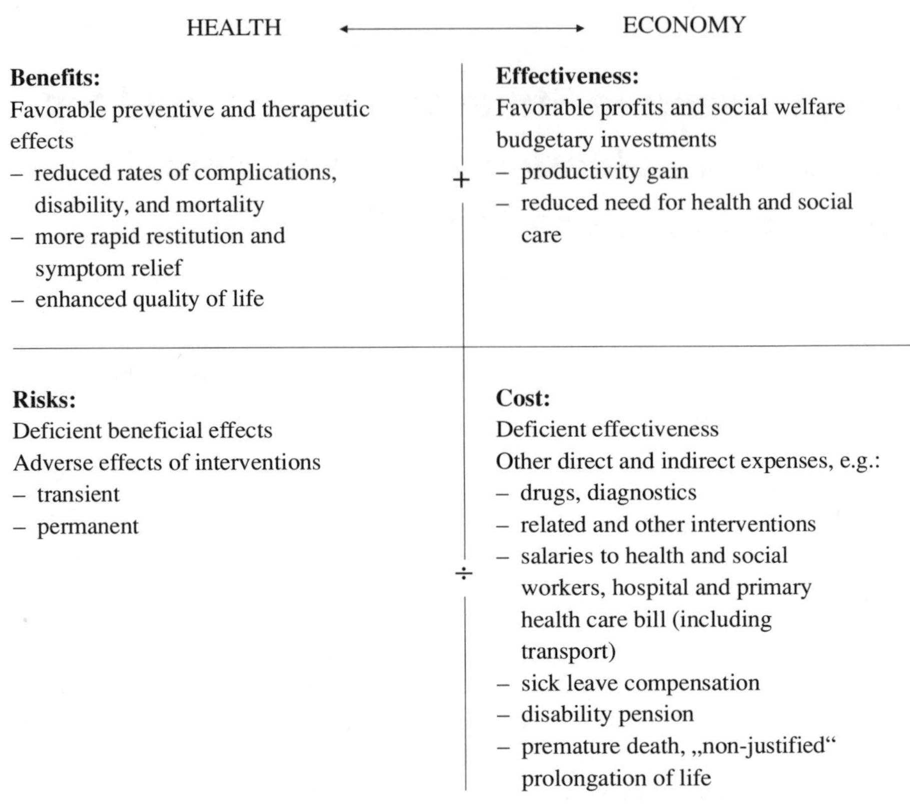

Fig. 1. Some elements to be weighted in a health-economic balance (see [6])

Design, data sources, approaches and methods in DUS

The design of DUS has to consider the type of users to be addressed, i.e., whether they are health administrators or health authorities, drug manufacturers, health professionals (academic and clinical), social scientists and economists, the media, and/or consumers. The documentation has to be sufficiently uniform to ensure appropriate communication between users, particularly with respect to the interpretation of data. Above all, the limitations of separate and often fragmentary studies must be thoroughly considered.

A common core of methodologies together with a wide range of data sources and analytical approaches have been adopted, described and discussed [4–13].

Basic data sources include appropriate drug, health and morbidity, demographic, and other relevant general statistics. These statistics must be supplemented by a wide range of study approaches to accommodate the type of problems being addressed. The complexity of drug-related problems often calls for analyses at the site of action (the patient, the disease, the doctor, the drug, the event level) rather than distant monitoring. A combination of epidemiological methods such as cohort and case-control studies, controlled clinical trials and non-intervening clinical observations, other types of PMS/databases related to therapeutic and adverse events, socio-anthropological studies to better understand perception/attitude/compliance problems, quality-of-life estimates, various clinical pharmacological approaches (drug kinetics, pharmacodynamics, genetics, etc.) and health economic evaluations, is necessary.

Whether prospective, sometimes intervention-based studies or retrospective studies are desirable has to be weighted against scientific needs, ethics, and feasibility. The aforementioned approaches require: (*i*) continuous or regular intermittent monitoring and surveillances (comprehensive sample/event studies, multiple record linkages); (*ii*) stand-by preparedness (set-up of databases, study prototype protocols); and (*iii*) separate ad hoc studies. It should be clearly emphasized from the outset whether the studies are intended to be descriptive, analytical, or problem/intervention-based, and if feedback to specific target groups is relevant.

May the risks exceed the benefits of drug interventions?

Stigmatized drugs

There are many historical examples of unacceptable iatrogenic drug complications (e.g. [14–15]). However, even among "clearcut" examples like arsenic, mercury, antimony, bismuth compounds, chloroform, thalidomide, clioquinols (iodochlorhydroxyquins), and phenacetin, some doubts about their discontinuance may persist if emotions are set aside and the situations examined calmly and rationally. Excessive marketing, inappropriate use, and easy access to more attractive alternatives tend to be more important determinants of drug discontinuance than medico-legal actions proscribing their dissemination and use.

Recent examples of discontinued non-steroidal anti-inflammatory drugs (NSAIDs) include phenylbutazone [15], benoxaprofen [16], and zomepirac [17]. DUS are challenged to justify a nearly worldwide ban on these drugs. Phenylbutazone and oxyphenebutazone were stigmatized as being the most toxic NSAIDs. This has been confirmed vis-à-vis agranulocytosis and aplastic anemia [18]; however, indomethacin and some other NSAIDs seem to present similar risks [19]. These risks are generally quite low when the drugs are used in an appropriate manner. However, benoxaprofen, which rapidly gained a major part of the NSAID market in the U. K. and some other countries due in part to unjustified promotional claims, caused severe (sometimes lethal) liver damage in a number of patients [17]. Another peculiar feature of benoxaprofen was its propensity to cause phototoxic skin reactions.

Zomepirac, launched mainly as an analgesic, may not have been fully recognized as an NSAID, for it came as somewhat of a surprise when a number of lethal idiosyncratic reactions occurred. The justification for zomepirac's permanent withdrawal may be harder to understand. A major reason may be that these adverse events occurred concomitantly with the stigmatization (rightly or wrongly) of several other NSAIDs.

Controversies in the area of drugs and therapeutics usually focus on specific drugs and drug products. This may have occasionally diverted drug control bodies, the health professions, the media, and the public away from more fundamental issues of therapeutics. There are many possible reasons for focussing on specific items such as minor differences between closely-related single drugs and drug products, uneven documentation of the excessive number of available drugs, and the often arbitrary adoption of widely-different drug therapy practices. Thus, balanced and qualified opinions are difficult to come by, as the consequences are only partly elucidated through DUS and related approaches.

To treat or not to treat – major areas of conflict and concern

The basic purposes for administering drugs are: (*a*) to cure diseases and their complications; (*b*) to prevent diseases and their complications; (*c*) to alleviate symptoms; and (*d*) to facilitate other interventions, e.g. surgery, psychosocial, and physical rehabilitation, etc. The available evidence indicates that the potential curing effects of drugs are limited, at least from a long-term perspective; anti-infectives and some cytostatics serve as relevant examples of this viewpoint. Thus, it should be recognized that drugs mainly serve purposes (*b*)–(*d*).

During the last 20–30 years, increasing efforts have been undertaken to document the clinical value and risks of drugs in relation to prevalent diseases such as cardiovascular disorders, peptic ulcer, and rheumatism and other arthritic disorders. Controlled clinical trials, often in the form of huge Phase IV studies, are adopted as the main tools. In some areas, such as thrombolytic therapy [20], impressive favorable effects have been obtained for reduced reinfarction and medium-term mortality. In other areas, like mild and moderate hypertension [21], the beneficial effects are marginal and some risks, especially related to acute myocardial infarction, may outweigh the benefits.

More recently, concern is increasing as to the relevance of controlled clinical trials when the results are extrapolated to everyday clinical practice. One shortcoming of these trials is that most of them are of much shorter duration than what may be the intention in clinical practice, so important clinical end-points may not be elucidated. Another confounding factor relating as well to major long-term trials is that the design (inclusion/ exclusion criteria, limited possibilities for individualized treatment, close control and intensive monitoring) may produce results which are not directly transferable to routine medical practice. Consequently, results from clinical practice might be more and less beneficial than those derived from controlled trials.

The following examples illustrate these dilemmas:

A. The mortality rate of patients who had suffered myocardial infarctions and were undergoing long-term treatment with several anti-arrhythmic drugs was more than double that of patients in a placebo group [22]. These findings should not have come as a complete surprise, as various re-evaluations of earlier studies indicated some risk of increased mortality for a number of anti-arrhythmic drugs [23,24]. Although selective categories of patients were studied (mainly those with ventricular arrhythmia in post-infarction situations), these findings call for more restricted clinical practice, especially so far as non-symptomatic arrhythmias are concerned. On the other hand, intermittent treatment may still be life-saving or, more importantly, may have positive effects on the quality of life of patients with severe symptomatic arrhythmias.

B. Tendencies to start drug treatment in mild to moderate hypertension have varied extensively over the years [4,25,26]. Likewise, the tendency to consider hypertension in the context of concomitant cardiovascular diseases and other risk factors has also been unclear. Disturbingly, the findings in a recent screening and follow-up study in four Norwegian counties indicated excess mortality from *all* causes in the treated groups [27]. This provocative study, which closely reflected clinical practice, also suggested that excess mortality is higher in mild and moderate hypertension than in more severe hypertension. Quality of life aspects obviously add to these dilemmas.

C. A recent meta-analysis of the six major anti-lipidemic drug and diet studies produced results similar to those mentioned in examples **A** and **B**. Except for one of the re-analysed studies, total mortality was at least as high in the intervention groups as in the control groups [28]. The mortality in the former was partly due to deaths of a violent nature, which might reflect psychiatric disturbances related to drug and dietary interventions. Again, quality of life indicators should be thoroughly considered.

None of the above examples can be considered final and definitive proofs of the unfavorable effects of these common interventions. However, they raise extremely important clinical and ethical questions, the responses to which also have health economic implications. Above all, they indicate the need for comprehensive DUS.

D. A somewhat different but nonetheless extremely challenging area is the present state of intervention approaches for peptic ulcer and similar disorders. Successful drug interventions during the last 15–20 years has dramatically reduced the need for surgery and sick leave, thereby reducing total costs for several of these disease entities [29]. However, the situation is not without problems such as extensive relapse rates even during continuous drug treatment and a potentially increased risk of acute complications [30]. The relapse rate seems to be lowered by using old drugs like bismuth compounds, preferably in combination with anti-infectives to eradicate *Helicobacter pylori*, which is associated with some forms of ulcer [31]. This approach, still in the experimental stage, may solve some problems while introducing others such as microbial resistance, additional gastrointestinal (GI) adverse effects, and bismuth toxicity.

Potential of NSAIDs: benefits and risks

No firm evidence presently exists to suggest that NSAIDs, unlike other drugs used for rheumatic disorders (gold, penicillamine, etc.) can modify the fundamental disease processes [32]. When NSAIDs are used in acute injuries, however, a slight reduction in recovery time may be noted. The main beneficial roles of NSAIDs are relief of painful stiffness, improved functional capacity within the limits set by the stage of disease, and facilitated rehabilitation through physiotherapy. There is no doubt that NSAIDs *can* contribute significantly to the quality of life.

The predominant indications for NSAIDs refer to continuous or long-term intermittent use in osteoarthritis and rheumatoid arthritis [33]. They are also used, though less frequently, for a wide range of musculo-skeletal disorders, for cancer pain relief, and for acute trauma. About two-thirds of these drugs are consumed by the elderly (>60–65 years of age).

The potential risks of NSAIDs have caused extensive and sometimes exaggerated concerns. Some 15–20% of the reports collected in spontaneous ADR reporting systems refer to NSAIDs [34]; most of these reactions are of a moderate or severe nature. According to controlled pre- and post-registration clinical trials [35] severe ADRs (most often of a GI nature, i.e. ulcers with subsequent perforations and bleeding [36–38]) occur in 1–2% of selected patient cohorts. The importance of drug formulations should also be emphasized [39]; in this context, the sustained-release indomethacin product Osmosin is a quite distinct example.

The ADR profile of various NSAIDs is qualitatively uniform and extends beyond GI complications to encompass idiosyncratic reactions, bone marrow depression (agranulocytosis, aplastic anemia [18,19]), hepatic damage, renal failure, electrolyte/fluid balance disturbances (sometimes with secondary hemodynamic implications), and central nervous system adverse experiences [40]. The relative incidence of such reactions is more variable [33,41], a matter which has been subject to extensive disputes at regulatory, drug industry, prescriber, and consumer levels.

Various rankings of risks according to type and severity of ADRs do not allow firm conclusions about the safety of NSAIDs, except that ibuprofen in the dosage presently recommended for over-the-counter (OTC) sale seems relatively safe, and that benoxaprofen and sulindac have a greater tendency to cause liver damage than most other NSAIDs. There is still some concern about NSAIDs with long half-lives, however. The relative risk of aspirin has partly been related to the type of formulation [39], but from a systemic point of view the relative risk is unclear. With considerable reservations one may tentatively conclude that the risk of any particular NSAID is, to some extent, related to its therapeutic potential and dosage.

During the early and mid-1980s much attention was focussed on the assumed excessive risk of one of the world's leading NSAIDs, Feldene (Pfizer's brand of piroxicam). One of its main competitors, Naprosyn (naproxen, manufactured by Syntex) had never been stigmatized to that extent. In Norway, Sweden, and several other countries a vigorous marketing campaign gave Feldene a 20–25% share of the NSAID market soon after registration (Fig. 2). This fact may not have been fully appreciated when an apparently excessive number of ADRs were received by ADR committees and substantiated clinically by gastroenterologists [42]. The ensuing uproar caused sales of Feldene to plummet dramatically. This led to Pfizer's undertaking a multicenter double-blind phase IV study to compare Naprosyn and Feldene vis-à-vis efficacy and safety in a cohort of 2,035 patients suffering from osteoarthritis, aged 17 years and upwards, recruited from the general population [35]. The treatment period was 12 weeks, and the initial doses were 750 mg and 20 mg for Naprosyn and Feldene, respectively. After 4 and 8 weeks control and monitoring took place, with the option that doses could then be reduced to 500 mg of Naprosyn and 10 mg of Feldene, according to the clinical situation. No major differences in efficacy or safety were observed between the two drugs in terms of pain relief or functional activity, and the 1% rate of moderate to severe ADRs found for the drugs corresponded to what had been seen in pre-registration trials. Surprisingly, an *inverse* correlation was found between adverse events and age.

This study, one of the largest comparative phase IV studies on NSAIDs ever performed, was subject to quite extensive criticism, not the least being in Norway [43]. A major criticism concerned the justification of performing such a large study to confirm what was in fact *already* broadly known. Nonetheless, the number of patients was far too small to be able to detect statistically meaningful differences

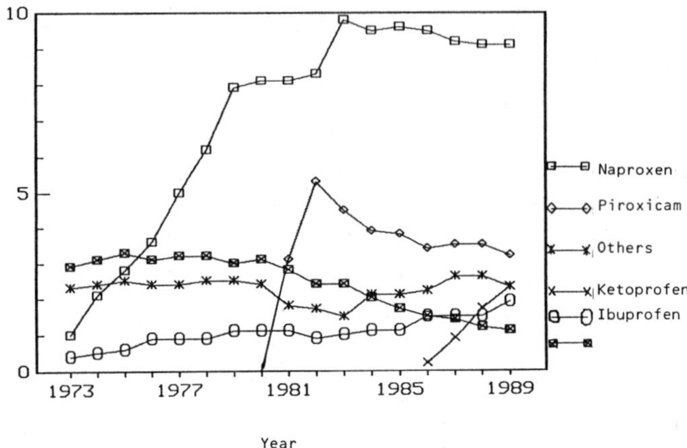

Fig. 2. Sales of NSAIDs and other anti-rheumatic drugs in Norway between 1973 and 1989, in Defined Daily Doses (DDD)/1000 inhabitants/day. "Others" refers to sulindac (0.9 DDD in 1989), gold, penicillamine, and chloroquine preparations. (From [45].)

in the ADR rate (1.0% vs. 1.5%) between the two drugs; to do so would have required 5,000–6,000 patients in each group, but still with some doubts about the equivalence of the doses compared.

This controlled study, due to the exclusion/inclusion criteria and other study design elements, is actually quite far removed from everyday clinical practice. One may still assume, however, that the apparently excessive ADR rate, at least during the early post-registration phase, was real, and was likely related to the frequency of use of the new drug; i.e., physicians are always tempted to use a new drug whenever patients have not tolerated or responded satisfactorily to other drugs within a therapeutic class. Nonetheless, the above study was probably of major importance to Pfizer which, during a subsequent FDA hearing, was able to save for itself a substantial share of the world NSAID market [44]. Needless to say, the producer of Naprosyn, Syntex, was somewhat less enthusiastic about this outcome [45].

As seen in Fig. 2, Feldene never fully regained its market share in Norway. In that country, total NSAID sales, excluding aspirin, have stabilized at around 20 Defined Daily Doses (DDD)/1,000 inhabitants/day, corresponding to a "continuous" exposure of some 2% of the population [46]. Aspirin use decreased by more than 50% in the period 1977–1989 (Fig. 3), reflecting recommendations to the public that paracetamol is a safer choice for minor pains and fever. Packages containing more than 20–30 tablets (about 10–15 g) of aspirin and paracetamol

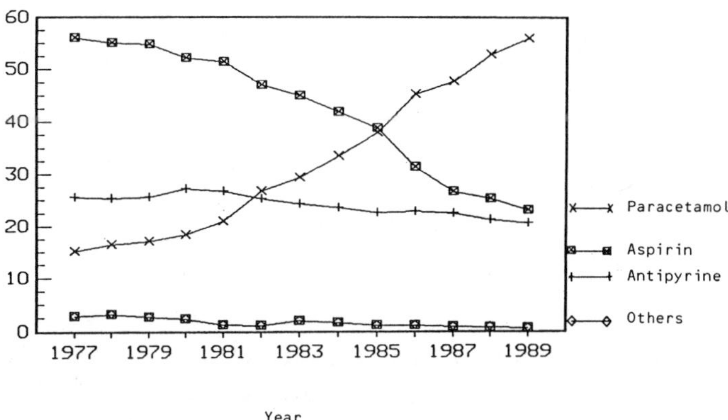

Fig. 3. Sales distribution (%) of antipyretic analgesics (ATC class N 02 B) in Norway between 1977 and 1989. (From [45].)

are now prescription drugs. Ibuprofen in smaller packages is allowed as an OTC drug, with the recommendation that daily doses for self-medication not exceed 800–1,000 mg.

Recent sales data on NSAIDs from the five Nordic countries (Tab. 1) show that sales figures for Norway remain remarkably stable and low compared to Denmark, Finland, Iceland, and Sweden [47]. The number of drug entities and relative preferences also vary widely from country to country.

General comments and conclusions

NSAIDs comprise a wide range of drugs, of which 10–12 are in wide use, and which are assumed to elicit their actions by influencing the prostaglandin cascade and various other biological systems [32]. The relevance of these mechanisms of action to the observed clinical effects is not fully known. A lack of fundamental understanding of rheumatic disease processes adds to the gap between what is ideally intended and what is feasible by drug and other interventions.

The separate evaluation of benefits and risks from drug interventions remains problematic [48]. Unpredictable disease and genetic factors may contribute positively or negatively to any interventions, including ADRs. The firm clinical experience of patient preference between alternative NSAIDs is very difficult to prove or disprove in controlled clinical studies, but should be viewed in the context of ADRs and dosage. Concomitant supplementary interventions also add to

30

Table 1. Statistics on NSAID sales and product registration in Nordic countries (from [45,46]).

	Year	Denmark	Finland	Iceland	Norway	Sweden
Total sales (DDD/1,000 inhab./day)	1987	22.9	32.4	29.8	18.8	21.0
	1988	23.3	35.4	30.1	19.6	23.4
	1989	24.7	42.8	32.5	20.0	25.9
No. of NSAIDs registered		15	12*	9	7	11

* Finland has by far the highest products/brands ratio

evaluation difficulties when groups of patients are compared. These aspects require more problem-oriented and comprehensive studies, and it is of utmost importance to design clinical observation studies reflecting everyday clinical practice as a supplement to controlled trials.

References

1. Strom, B.L. (Ed.), Pharmacoepidemiology – the science of postmarketing surveillance. Churchill Livingstone, New York 1989.
2. Lunde, P.K.M. and Tognoni, G., Definitions relevant to drug utilization. In: World Health Organization Drug Utilization Research Group "EURO", *Drug Utilization studies: methods and uses*. World Health Organization, Copenhagen (1991, in press).
3. World Health Organization, The selection of essential drugs. World Health Organization Technical Report Series no. 615. World Health Organization, Geneva 1977.
4. Sjöqvist, F. and Agenäs, I. (Eds.), Drug utilization studies: implications for medical care. Acta Med. Scand., suppl. 683: 1–152 (1984).
5. Baksaas, I. and Lunde, P.K.M., National drug policies: the need for drug utilization studies. Trends Pharmacol. Sci. **7**: 331–334 (1986).
6. Lunde, P.K.M. and Agenäs, I. (Eds.), Drug utilization in relation to morbidity. Acta Med. Scand., suppl. 721, 1–35 (1988).
7. Lunde, P.K.M., Differences in national drug-prescribing patterns. Clinical pharmacological evaluation in drug control. Report on a World Health Organization Symposium, Deidesheim, Germany, 11–14 November 1975. World Health Organization Regional Office for Europe, Copenhagen 1976, pp. 19–47.
8. Bergman, U., Grimsson, A., Wahba, A.H.W., and Westerholm, B. (Eds.), Studies in drug utilization: methods and applications. World Health Organization Regional Office for Europe, Publication Ser. no. 8. World Health Organization, Copenhagen 1979.
9. Lunde, P.K.M., Levy, M. et al., Drug utilization: geographical differences and clinical implications. In: Duchene-Marullaz, P. (Ed.), Clinical pharmacology, vol. 6: Advances in pharmacology and therapeutics (Proc. 7th Int. Congr. Pharmacol., Paris 1978). Pergamon Press, Oxford 1979, pp. 77–170.
10. Serradell, J., Bjornson, D.C., and Hartzema, A.G., Drug utilization study methodologies: national and international perspectives. Drug Intell. Clin. Pharm. **21**: 994–1001 (1987).

11. Lee, D. and Bergman, U., Studies of drug utilization. In: Strom, B.L. (Ed.), Pharmacoepidemiology. Churchill Livingstone, New York 1989, pp. 259–274.

12. World Health Organization Collaborating Centre for Drug Statistics Methodology, Drug utilization bibliography 1981–1989. World Health Organization, Copenhagen 1989.

13. World Health Organization Drug Utilization Research Group "EURO", Drug Utilization studies: methods and uses. World Health Organization, Copenhagen 1991 (in press).

14. Soda, T. (Ed.), Drug-induced sufferings: medical, pharmaceutical and legal aspects (International Congress Series no. 513). Excerpta Medica, Amsterdam 1980.

15. Penn, R.G., Iatrogenic disease: an historical survey of adverse reactions before thalidomide. In: D'Arcy, P.F. and Griffin, J.P. (Eds.), Iatrogenic diseases, 3rd edn. Oxford University Press, Oxford 1986, pp. 14–21.

16. Editorial, Benoxaprofen. Br. Med. J. **285**: 459 (1982).

17. del Favero, A., Anti-inflammatory analgesics used in rheumatoid arthritis and gout. In: Dukes, M. N. G. (Ed.), Side effects of drugs annual, no. 8. Elsevier, Amsterdam 1984, pp. 100–117.

18. The International Agranulocytosis and Aplastic Anemia Study (IAAAS), Risks of agranulocytosis and aplastic anemia: a flrst report of their relation to drug use with special reference to analgesics. J.Am. Med.Assn. **256**: 1749–1757 (1986).

19. The International Agranulocytosis and Aplastic Anemia Study (IAAAS), Study Monograph. Oxford University Press, Oxford 1991 (in press).

20. Gruppo Italiano per lo Studio della Sopravivenza nell'Infarto Miocardico, GISSI-2: A factorial randomised trial of alteplase versus streptokinase and heparin versus no heparin among 12 490 patients with acute myocardial infarction. Lancet **336**: 65–75 (1990).

21. Medical Research Council Working Party, MRC trial of treatment of mild hypertension: principal results. Br. Med. J. **291**: 97–104 (1985).

22. The Cardiac Arrhythmia Suppression Trial (CAST) Investigators, Preliminary report: effect of encainide and flecainide on mortality in a randomized trial of arrhythmia suppression after myocardial infarction. N. Engl. J. Med. **321**: 406–412 (1989).

23. Anderson, J.L., Should complex ventricular arrhythmias in patients with congestive heart failure be treated? A protagonist's viewpoint. Am. J. Cardiol. **66**: 447–450 (1990).

24. Podrid, P.J. and Wilson, J.S., Should asymptomatic ventricular arrhythmia in patients with congestive heart failure be treated? An antagonist's viewpoint. Am. J. Cardiol. **66**: 451–457 (1990).

25. Lunde, P.K.M., Andrew, M., Baksaas, I., Drug utilization studies–an instrument in drug research. In: Kewitz, H., Roots, I., and Voigt, K. (Eds.), Epidemiological concepts in clinical pharmacology. Springer Verlag, Berlin 1987, pp. 57–72.

26. Hense, H.W. and Tennis, P., Changing patterns of antihypertensive drug use in a German population between 1984 and 1987. Results of a population based cohort study in the Federal Republic of Germany. Eur. J. Clin. Pharmacol. **39**: 1–7 (1990).

27. Thürmer, H.L., Lund-Larsen, P.G., Tverdal, A.A., and Thelle, D.S., Treatment of hypertension as a risk factor in a prospective study. J. Risk Safety Med. **1**: 267–278 (1990).

28. Muldoon, M.F., Manuck, S.B., and Matthews, K.A., Lowering cholesterol concentration and mortality: A quantitative review of primary prevention trials. Br. Med. J. **301**: 309–314 (1990).

29. Pym, B., Sandstad, J., Seville, P., Byth, K., Middleton, W.R.J., Talley, N.J., and Piper, D.W., Cost-effectiveness of cimetidine maintenance therapy in chronic gastric and duodenal ulcer. Gastroenterology **99**: 27–35 (1990).

30. Goodwin, C.S., Duodenal ulcer, *Campylobacter pylori*, and the "leaking roof" concept. Lancet **331**: 1467–1469 (1988).

31. Sidebotham, R.L. and Baron, J.H. Hypothesis: *Helicobacter pylori,* urease, mucus, and gastric ulcer. Lancet **335**: 193–195 (1990).

32. Insel, P., Analgesics–antipyretics and antiinflammatory agents: drugs employed in the treatment of rheumatoid arthritis and gout. In: Gilman, A.G., Rall, T.W., Nies, A.S., and Taylor, P. (Eds.), Goodman and Gilman's The pharmacological basis of therapeutics, 8th edn. Pergamon Press, New York 1990, pp. 638–681.

33. Drug Surveillance Research Unit, Comparative study of five NSAIDs. Prescript. Event Monitor. News **3**: 3–12 (1985).

34. Kromann-Andersen, H., Kovacs, I. and Pedersen, A., ADRs during use of NSAIDs in Denmark during a period of 15 years. Ugeskr. Laeger **148**: 462–468 (1986).

35. Husby, G., Holme, I., Rugstad, H.E., Herland, O.B., and Giercksky, K.E., A double-blind multicentre trial of piroxicam and naproxen in osteoarthritis. Clin. Rheumatol. **5**: 84–91 (1986).

36. Somerville, K., Faulkner, G., and Langman, M., Non-steroidal anti-inflammatory drugs and bleeding peptic ulcer. Lancet **I**, 462–464 (1986).

37. Walt, R., Katschinski, B., Logan, R., Ashley, J., and Langman, M., Rising frequency of ulcer perforation in elderly people in the United Kingdom. Lancet **I**, 489–492 (1986).

38. Faulkner, G., Prichard, P., Somerville, K., and Langman, M.J.S., Aspirin and bleeding peptic ulcers in the elderly. Br. Med. J. **297**: 1311–1313 (1988).

39. Broers, 0., Gastrointestinal mucosal lesions: a drug formulation problem. Med. Toxicol. **2**: 105–111 (1987).

40. Dukes, M.N.G. (Ed.), Meyler's side effects of drugs, 11th edn. Elsevier, Amsterdam 1988.

41. Anonymous, Which NSAID? Drug Ther. Bull. **25**: 81–84 (1987).

42. Laake, K., Kjeldaas, L., and Borchgrevink, C.F., Side effects of piroxicam (Feldene). Acta Med. Scand. **215**: 81–83 (1984).

43. Borchgrevink, C., Drug treatment of arthrosis. Tidsskr. Nor. Laegeforen. **106**: 2485–2486 (1986).

44. Anonymous, Syntex on Feldene/Naprosyn. Scrip no 1090/91, 18 (1986).

45. Anonymous, Piroxicam vs. naproxen assessed. Scrip no. 1076, 26 (1986).

46. Ullerud, T.G. and Sakshaug, S. (Eds.), Drug use in Norway. Norwegian Medicinal Depot, Oslo 1990.

47. Nordic Statistics on Medicines 1987–1989 (NLN publication no. 30). Nordic Council on Medicines, Uppsala 1990.

48. Kåss, E., Drugs used in rheumatic diseases. In: Sakshaug, S. et. al (Eds.), Drug utilization in Norway during the 1970s: increases, inequalities, innovations. Norwegian Medicinal Depot, Oslo 1983, pp. 163–175.

Adverse Reactions to NSAID: Clinical Pharmacoepidemiology
ed. by M. Kurowski
© 1992 Birkhäuser Verlag Basel

Post-Marketing Surveillance as an Instrument of Adverse Drug Reaction Monitoring

H. Kewitz

Institute of Clinical Pharmacology, Klinikum Steglitz der Freien Universität Berlin, Hindenburgdamm 30, D-W-1000 Berlin 45, Germany

Post-marketing surveillance (PMS) of drugs involves a number of procedures for observing and registering the benefits and risks of drugs following their approval for general use. Unfortunately, in Germany most PMS activities are sporadic, non-coordinated, and restricted to the short term, and spontaneous reporting of adverse drug reactions (ADRs) remains the only continuously-maintained PMS procedure since it was instituted in 1961 by Gertrud Hohmann, then secretary of the Arzneimittelkommission der Deutschen Ärzteschaft (Commission on Drugs of the German Physician's Association [1]).

In Berlin in 1971 we undertook a PMS project with Werner Altwein which led us to join the Boston Collaborative Drug Surveillance Program a year later [2]. In this report, some of our data on analgesics and non-steroidal anti-inflammatory drugs (NSAIDs) collected between 1971 and 1980 are presented. These data are of two types: those on hospital admissions due to ADRs to analgesics and NSAIDs, and those on ADRs to the same drugs occurring in-hospital.

The first set of data is an expansion of that published in collaboration with Professor M. Levy in 1980 on ADR hospital admissions in Jerusalem and Berlin [4].

In the Boston Collaborative Drug Surveillance Program, data were collected on all consecutive admissions without further selection. Patients were questioned about the medications they used during the 4 weeks prior to admission, and the admitting doctor was asked whether the present admission could have been caused by an ADR. The doctor's answer to this question, including the implicated drug, was considered a suggestion to be critically assessed on the basis of that physician's experience, professional education, and knowledge and awareness of ADRs from the medical literature and other sources. One-sided views may have been "smoothed out" by the large number of attending physicians who shared the duty of admitting patients. The reactions described were regarded as the major but not necessarily the sole cause of hospitalization. The data reflect neither relative nor absolute risks, nor are they to be construed as generally or universally representative; rather, they are an indication of the magnitude, severity, and type of drug-induced diseases occurring in the community of Berlin-Steglitz during the years 1975–1980. It is noteworthy, however, that there is an astonishing conformity between the data from Berlin-Steglitz and those obtained at other acute-care hospitals in other parts of the world.

The usual rate of ADR admissions ranges from 2% to 7%; in our hospital it was 6%. The rate of admissions due to adverse reactions (ARs) to analgesics or NSAIDs was 74 out of a total of 6,000 admissions, corresponding to 1.2% of all admissions and almost 20% of all ADR admissions (Tab. 1). This is commensurate with the heavy use of such drugs in the community. Fourteen percent of the admitted patients had used medications for headache (principally aspirin) during the preceding 4 weeks, and 17% had used mostly non-aspirin NSAIDs. These high exposure rates demonstrate that many people are willing to take the chance of relieving pain by taking one or two tablets which they consider to be relatively harmless. Indeed, these drugs rarely produce life-threatening or even serious ADRs. Due to the low incidence of adverse events and the frequent use of drug combinations, quantitative comparisons of different compounds are difficult to analyze.

Table 1. Boston Collaborative Drug Surveillance Program at Berlin-Steglitz, 1975–1980.

	No. of Patients	% of Total
Total	6000	100.0
Patients using analgesics for headache (4 weeks before admission)	834	13.9
Patients using NSAIDs (4 weeks before admission)	1022	17.0
Admissions for ADRs to analgesics and NSAIDs	74	1.2

Table 2 gives an idea of the confounding factors introduced by concomitant use of two or more analgesics and NSAIDs. With respect to ADR admissions, aspirin was the drug with the highest exposure rate, closely followed by pyrazolones (e.g., dipyrone, or metamizol); exposures to phenacetin (which has been gradually replaced by paracetamol) and phenylbutazone (which has been subject to ever-increasing legal restrictions) were considerably lower. Use of phenacetin and phenylbutazone has been largely supplemented by indomethacin, diclofenac, and ibuprofen, which are arylacetic acid/arylpropionic acid derivatives.

Table 2. Hospital admissions due to ADRs to analgesics and NSAIDs.

	No. of Patients
Total	74
Drug exposure 4 weeks prior to hospital admission:*	
Aspirin	44
Pyrazolones	36
Phenacetin/paracetamol	19
Phenylbutazone	14
Arylacetic acids/arylpropionic acids	12

* some patients concomitantly took more than one type of drug, hence a per-drug total >74.

Table 3 shows a breakdown by symptoms and signs of ADR hospital admissions in relation to the implicated drugs. The most frequently-encountered serious ADR requiring hospitalization was major gastrointestinal (GI) bleeding, which has a mortality of almost 10%. In about half the cases of GI bleeding aspirin was considered the causal drug, though it was very often used in combination with other analgesics, particularly paracetamol and/or pyrazolones. Phenylbutazone without aspirin was associated with a somewhat lower incidence of GI bleeding, and indomethacin was associated with a still lower incidence of GI bleeding.

Curiously, thrombocytopenia was often considered to be caused by aspirin or aspirin in fixed combinations, whereas pyrazolones were suspected of causing thrombocytopenia in only one case. Two cases of aplastic anemia were associated with the use of phenylbutazone, and in one case diclofenac was thought to be the causal drug. Aplastic anemia is certainly one of the most worrisome drug-induced diseases, since it is almost irreversible and has a mortality rate of 50% within 2 years of onset [5].

In contrast, agranulocytosis has a much better prognosis, patients having a good chance of full recovery without sequelae within one or two weeks. Acute mortality for agranulocytosis is around 10%, especially for those cases which are not diagnosed early enough and the responsible drug is not discontinued [5].

Hemolytic anemia was probably caused by an aspirin combination in one case and by a pyrazolone combination in another. Both patients were concomitantly exposed to aspirin and pyrazolones, again totally confounding the analysis.

A number of cases of nephropathy caused by chronic and/or heavy use of analgesics are diagnosed every year. In our study we encountered five such cases, three due to phenacetin and two due to paracetamol, all in fixed combinations with aspirin and caffeine. Therefore, the effects of phenacetin and paracetamol vis-à-vis nephropathy were totally confounded by both aspirin and caffeine. Other investi-

Table 3. Symptoms leading to ADR hospital admissions and the implicated drugs.

Reaction	Drug	No. of Patients
Gastrointestinal bleeding	Aspirin Aspirin combinations	5 20
	Phenylbutazone Phenylbutazone combinations	3 7
	Indomethacin Indomethacin combinations	4 1
	Pyrazolones	3
	Arylpropionic acids	2
	Total	45
Thrombocytopenia, abnormal coagulation	Aspirin Aspirin combinations	3 7
	Pyrazolone	1
	Total	11
Aplastic anemia, pancytopenia	Phenylbutazone Phenylbutazone combinations	1 1
	Diclofenac (arylacetic acid)	1
	Total	3
Agranulocytosis, leucopenia	Pyrazolone Pyrazolone combinations	1 1
	Total	2
Nephropathy	Phenacetin combinations Paracetamol combinations	3 2
	Total	5
Allergic reactions	Aspirin combinations Pyrazolone Pyrazolone combinations	1 1 2
	Total	4
Vertigo, tinnitus	Aspirin combinations	2
Hemolytic anemia	Aspirin combinations Pyrazolone combinations	1 1
	Total	2

gators had the same problem with cases of drug-induced nephropathy, and it is not yet clear whether the monopreparation or the combination with either aspirin and/or caffeine are required to cause renal damage.

In 4 cases of allergic reactions involving intense urticaria and fever, 3 were probably caused by pyrazolones and 1 by aspirin, though the pyrazolones were formulated as combinations containing aspirin. Tinnitus and severe vertigo appeared in two cases; in both cases aspirin was the suspected compound.

Table 4 summarizes ADR hospital admissions according to drug type and adverse event. It is noteworthy that 50% of the ARs were attributable to aspirin, whereas only 13% were attributed to pyrazolones. These two kinds of analgesics are generally considered equipotent with regard to pain and fever remission. Phenylbutazone and indomethacin have similar indications and shared the same risks: both were implicated in causing GI bleeding and aplastic anemia. It is likely that exposure rates for phenylbutazone and indomethacin were not very different, albeit much lower (perhaps one-tenth) than for aspirin or pyrazolones. However, phenylbutazone and especially indomethacin were usually prescribed on a longterm basis for the treatment of chronic conditions.

All of the ARs described above required hospitalization for periods generally ranging from one to three weeks (Tab. 5). Patients with agranulocytosis or vertigo stayed in hospital for 1 week, whereas those with aplastic anemia or nephropathy

Table 4. Analgesics and NSAIDs implicated in ADR hospital admissions.

Drug	Adverse Reaction	No. of Patients
Aspirin Aspirin combinations	GI bleeding Thrombocytopenia, abnormal coagulation Allergic reactions Vertigo, tinnitus Hemolytic anemia	25 10 1 2 1
Phenylbutazone Phenylbutazone combinations	GI bleeding Aplastic anemia, pancytopenia	10 2
Pyrazolones Pyrazolone combinations	GI bleeding Thrombocytopenia, abnormal coagulation Agranulocytosis, leucopenia Allergic reactions Hemolytic anemia	3 1 2 3 1
Indomethacin and other Arylacetic/arylpropionic acids	GI bleeding Aplastic anemia	7 1
Phenacetin combinations Paracetamol combinations	Nephropathy	5
Total		74

Table 5. ADR hospital admissions and days of hospital treatment.

Reaction	No. of Patients	days in hospital (mean)	min.-max.
Gastrointestinal bleeding	45	16	6–45
Thrombocytopenia	11	16	9–30
Nephropathy	5	22	13–30
Aplastic anemia, pancytopenia	3	17	10–21
Agranulocytosis, leucopenia	2	6	4– 8
Allergic reactions	4	14	4–30
Vertigo, tinnitus	2	8	8– 8
Hemolytic anemia	2	12	11–13

needed hospital care for 3 weeks. In cases of aplastic anemia and nephropathy, however, many further hospital admissions can be expected over the ensuing months or years, with the life expectancy of these patients probably being considerably reduced.

Although the above data on ADR hospital admissions are not *sensu stricto* suitable for formal statistical analysis, they may in fact be extremely valuable in understanding to what extent therapeutically-useful drugs can cause serious health problems in the community. An ADR hospital admission rate of 6% is on a par with those values for diabetes mellitus or pulmonary infection. Such an admission rate is remarkable not only in terms of individual and public health but also in terms of health costs and patient care capacities. If, based on our data, one assumes 15 hospital admissions per year due to analgesic and NSAID ADRs, and only half of the ADR cases are properly recognized as such, then there would be ca. 300 ADR admissions per annum in the former West Berlin area; for a population of 2.1 million inhabitants, the annual incidence rate would be 300/2,100,000 or 3 per 21,000.

A different but nonetheless pertinent phenomenon is ADRs in patients treated with analgesics or NSAIDs who are *already* hospitalized for other, unrelated ailments. In this study, which followed a multiarmed cohort design, each admitted patient was monitored daily and every AR *to any drug* recorded by a specially-trained nurse. Table 6 contains data on four drugs which had likewise been examined vis-à-vis ADR hospital admissions: aspirin, metamizol, indomethacin, and phenylbutazone. While some "normalization" for sex, age, indication, and diagnosis would have been appropriate, it is still useful to directly compare the ADR rates for these drugs. As in the case of ADR hospital admission data, ARs to analgesics and NSAIDs by *already-hospitalized* patients were most often

Table 6. ADRs to analgesics and NSAIDs during in-hospital care*.

Adverse reaction	No. of ADRs attributable to:			
	Aspirin	Metamizol	Indo-methacin	Phenyl-butazone
Gastrointestinal bleeding	11	6	16	3
Sedation, tinnitus, vertigo	12	2	7	0
Thrombocytopenia	2	4	0	0
Leucopenia	0	1	0	0
Allergic reactions	0	7	1	1
Hypotension	1	0	0	1
Total	26	20	24	5
No. of ARs	425	1,956	257	101
Total/No. of ARs ($\times$ 100%)	6.1%	1.0%	9.3%	5.0%

*1975–1980: 6,000 patients

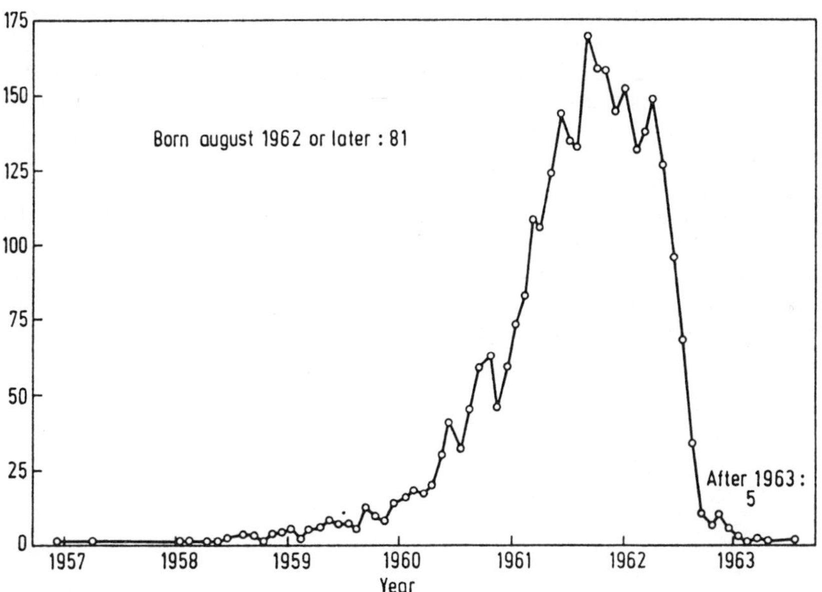

Fig. 1. Date of birth of 3049 victims of thalidomide embryophaty born in the Federal Republic of Germany.

manifested as GI tract disturbances, except for metamizol where allergic reactions were more common ADRs. Central nervous system impairment rather than thrombocytopenia was the next most frequent AR after GI disturbances for aspirin and indomethacin, for metamizol thrombocytopenia and leucopenia were the next most common ARs after allergic reactions. These data supplement and confirm our earlier observations relating to ADR hospital admissions.

Unfortunately the PMS described in this report has been terminated and has *not* been replaced by any other PMS programs anywhere in Germany.

A final point concerns the early detection of serious unforeseeable drug-induced diseases. Fig. 1 (from [3]) shows the incidence of phokomelia over some years. A similar situation, in which so many cases of a physiological disturbance occurred before its causal connection to drug use was identified, must not be allowed to recur. At present, nearly 30 years after the thalidomide disaster, we still do not have any PMS to quickly alert practitioners and the public of possible drug-induced maladies; in fact, there is not even any widespread support for developing such an instrument. Post-marketing drug surveillance is urgently needed in Germany on a formal and coordinated basis. Regrettably, it would appear that our society is not yet ready to support such work.

References

1. Hohmann, G., Die Erfassung und Vermeidung von Arzneimittelschäden. 20jährige Erfahrung in der Arzneikommission der Deutschen Ärzteschaft. Internist **94**: 19–24 (1973).
2. Jick, H., Comprehensive drug surveillance. J. Am. Med. Assn. **213**: 1455–1460 (1970).
3. Lenz, W., The Thalidomide hypothesis: how it was found and tested. In: Kewitz, H., Roots, I., and Voight, K. (Eds.), *Epidemiological concepts in clinical pharmacology*, Springer Verlag, Berlin 1987, pp. 3–10.
4. Levy, M., Kewitz, H., Altwein, W., Hildebrand, J., and Eliakin, M., Hospital admissions due to adverse drug reactions: a comparative study from Jerusalem and Berlin. Eur. J. Clin. Pharmacol. **17**: 25–31 (1980).
5. The International Agranulocytosis and Aplastic Anemia Study, Risks of agranulocytosis and aplastic anemia: a first report of their relation to drug use, with special reference to analgesics, J. Am. Med. Assn. **256**: 1749–1757 (1986).

Non-Steroidal Anti-Inflammatory Drugs: Epidemiological Studies of Adverse Reactions

M. Levy

Department of Medicine, Hadassah University Hospital, IL–91120 Jerusalem, Israel

This article is concerned with the analytical epidemiological approach to the evaluation of adverse drug reactions (ADRs), as exemplified by studies relating to non-steroidal anti-inflammatory drugs (NSAIDs). The advantage of analytical epidemiological studies of ADRs lies in their capability to quantify risk without evaluating causality for individual cases.

Methods Used for the Analytical Epidemiological Evaluation of Adverse Drug Reactions

Experimental Studies

Experimental studies, i.e., randomized controlled clinical trials, are a regular part of the pre-marketing phase of pharmaceutical development and are usually of limited clinical application, size, and duration. Detection of severe ADRs during the pre-marketing stage normally leads to discontinuance of the drug. Most of the data gathered during clinical trials of NSAIDs concern mild gastrointestinal (GI) disturbances. Since subjects must be randomly assigned to treatment and followup groups, experimental studies of ADRs in the post-marketing phase are difficult. Nonetheless, experimental studies should be considered in special circumstances.

Non-Experimental (Observational) Studies

In non-experimental studies there is no manipulation of drug exposure; whatever occurs is observed and recorded. Required data can be obtained from epidemiological surveillance systems monitoring drug exposures and ADRs as they occur in patient populations [1]. Another approach is to use record linkage systems of

42

computerized databanks, such as those containing information on drug prescriptions and registries of patient diagnoses [2]. Such multipurpose data systems are useful either to test or generate causal hypotheses, as, for example, in the study of NSAID use and GI bleeding. So far, however, most of the non-experimental studies have required the organization of ad hoc data collection systems.

A. Cohort studies

Cohort studies compare patients exposed to a drug with those not so exposed vis-à-vis events following exposure. The advantage of cohort studies is that during follow-up all events can be recorded and evaluated regardless of *a priori* suspicions as to whether they are in fact drug-related. If long-term follow-up is maintained, cohort studies can also be used to detect late effects. Cohort studies are costly and therefore applicable only to relatively common ADRs such as venous thromboembolism following oral contraceptive use [3]. They can be used to study the common (mainly non-specific) GI complaints reportedly experienced in up to one-third of NSAID users [4]. For the serious but rare ADRs to NSAIDs, cohort studies are impractical.

B. Case-Control studies

In contrast to cohort studies, case-control studies are useful for studying rare events. Cases with specific well-defined ADRs are compared to controls without the ADR vis-à-vis rates of drug use and other characteristics. Case-control studies enable evaluation of multiple causes of a particular ADR, which is an obvious advantage for the study of NSAID-related adverse events. The efficiency of case-control studies decreases when exposure is rare. Case-control studies have become the main source of most current epidemiological research on the causes of disease [5,6].

The methodological issues relevant to the design of case-control studies are based on common scientific principles: rigorous definition of the event to be studied and of the criteria for measuring of drug exposure (taking into account the etiologically-relevant period), systematic ascertainment of cases, proper selection of controls, and consideration of validity.

C. The quantitative estimation of risk

The basic estimate of an association between drug exposure and adverse event is the *relative risk* or *rate ratio*, defined as the ratio of the rate of occurrence of an event in those exposed to a drug (E_r) to the rate of occurrence in those not exposed to that drug (N_r) [7]. In cohort studies, where rates of disease are calculated directly among exposed and non-exposed comparison groups, the relative risk is derived by dividing the rates, i.e., E_r/N_r. In case-control studies the rates obtained are of exposure among those with and without the event under study. The relative risk is estimated by the odds ratio. The relative risk provides information about the stength of association between a drug and an adverse event.

A more important measure of risk is the *excess risk*, defined as the total incidence of an event among users of a drug minus the baseline incidence. The excess risk provides information about the number of cases to be expected among drug users as a consequence of exposure to that drug. For rare conditions (e.g. aplastic anemia) the relative risk can be quite misleading, as large rate ratios translate into low excess risks; on the other hand, for common diseases low relative risks translate into large excess risks.

Risk has also to be measured in terms of time. An etiologically-relevant period ("exposure window") has to be determined, which in turn depends on the kinetic properties of the event and of the drug.

In the evaluation of risk, the confounding effects of other drugs as well as non-drug factors have to be controlled.

Studies of ADRs to NSAIDs

It is their anti-inflammatory and analgesic properties which make NSAIDs the most commonly used drugs; in 1985 the worldwide market for NSAIDs was estimated as US$ 2.1 billion [4]. The most common adverse effects associated with NSAIDs are GI disturbances, particularly GI bleeding and perforation and gastric ulcers. Also of special concern with NSAIDs are nephrotoxicity, hepatotoxicity (including Reye's syndrome), blood dyscrasias, hemorrhagic diathesis, and anaphylactic and anaphylactoid skin reactions [8–10].

Major ADRs to mild analgesics – unless these drugs are abused or taken in high doses – are rare, and the risk to the individual is small. Nevertheless, because of the massive and universal use of these drugs, quantitative measurement of the risk involved (i.e. the public health hazard) is imperative. It is only by conducting epidemiological studies that data leading to decisions based on scientifically sound judgements can be obtained [11]

Analytical Epidemiological Studies of GI Reactions to NSAIDs

All NSAIDs are known to cause adverse GI reactions. Epidemiological studies relate to issues of gastric and duodenal ulceration, and GI bleeding and perforation.

44

A. Aspirin and GI bleeding

That aspirin causes major upper GI bleeding was first suggested many years ago by Douthwaite [12]. In the 1950s and 1960s a number of case-control studies were published, all showing a positive association between major upper GI bleeding and aspirin use. However, methodological deficiencies with regard to selection of cases and controls, the definition of exposure, and the control of confounding cast serious doubts on the validity of these studies. As is often the case, it could not be distinguished whether aspirin was taken before the bleeding commenced or whether the condition that caused the bleeding provoked aspirin use [13].

In 1974, a case-control study based on hospital admissions in the Boston area identified newly-diagnosed cases with gastric ulcer, duodenal ulcer, and upper GI bleeding with no other predisposing condition [14]. To ensure that aspirin use antedated the onset of illness, only regular aspirin use of three months' duration was analyzed. Heavy use (i.e. 4 days/week) was associated with bleeding (rate ratio, 2.1) and benign gastric ulcer (rate ratio, 3.4), whereas for duodenal ulcer or with less regular aspirin use there was no evidence for an association. An attempt to quantify the attributable (excess) risk for heavy, regular aspirin use produced estimates of 15 cases of massive upper GI bleeding and 10 cases of gastric ulcer per 100,000 users per annum. It is now recognized that these estimates are probably too low.

Case ascertainment was incomplete in this case-control study. Short and irregular aspirin use was included in the reference category, and aspirin use was defined in relation to hospital admission date rather than to the onset of symptoms. In 1983 Coggon et al. [15] compared aspirin and paracetamol consumption in matched pairs of patients experiencing major upper GI bleeding and community controls. Higher rates of drug use were found for the patients with GI bleeding. However, for aspirin an association was found between drug use and GI bleeding for both recent and habitual use, whereas for paracetamol an association was evident only for recent use, which could have been related to symptoms of the disease. It was indirectly estimated (based on hospital admission rates for major upper GI bleeding of 1 per 2,000–2,500 per year for the general population) that the excess risk for regular aspirin users amounts to ca. 40 per 100,000 per year, and that the risk of being admitted to hospital for major upper GI bleeding was 1 per 250,000 aspirin doses.

The obvious point that drug histories have to antecede the day on which symptoms commence rather than the day of hospital admission was considered in two recent case-control studies [16,17]. The risk of first episode of major upper GI bleeding in subjects not known to be predisposed to this problem was assessed in relation to the use of NSAIDs. In the study of Levy et al. [16], for aspirin use for at least four days within the week before the onset of symptoms the rate ratio estimate was 15 (lower 95% confidence limit (c.l.), 6.4), while for occasional use the rate ratio estimate was 5.6 (c.l., 2.7). In Kaufman et al.'s study [17], the overall rates were smaller and seemed to be dose-related; however, even low doses of aspirin (such as those taken for prophylaxis against myocardial infarction) appeared to be associated with GI bleeding. While the methods used in these two studies do not allow estimation of excess risk, the results suggest that the risk of GI bleeding to aspirin users is substantially higher than previously thought. In another case-control study it was found that elderly persons who had taken aspirin were two to three times more likely to be admitted to hospital with bleeding ulcers than elderly persons not taking aspirin. When all NSAIDs were considered, it was suggested that those drugs were responsible for over a third of admissions for bleeding peptic ulcers in the elderly [18].

Further support for the association between aspirin and GI bleeding came from post-myocardial infarction cohort studies, in which excessive GI bleeding was found in the aspirin-treated group, including those who took 325 mg every other day, compared to those treated with a placebo [19–21].

B. Non-aspirin NSAIDs (NANSAIDs) and upper GI bleeding

In the hospital-based control study described earlier [16], the risk of a first episode of major upper GI tract bleeding in subjects not known to be predisposed to this ailment was also assessed in relation to the use of NANSAIDs. For regular NANSAID use during the week before onset of symptoms, the adjusted rate ratio estimate was 9.1 (c.l., 2.7–3.1); there were insufficient data to evaluate individual drugs or occasional use. Other case-control studies of GI bleeding and NANSAIDs have also shown a significant association, although the estimates of relative risk were somewhat lower [22–24]. Elderly patients receiving NANSAIDs are more likely to develop GI complications. NANSAIDs were taken more than twice as often in patients with small and large bowel perforation and hemorrhage compared to controls [25].

The cohort approach for the study of GI bleeding in relation to NANSAIDs was attempted using the prescription-event monitoring system of the drug surveillance research unit of the University of Southampton (U. K.) and other data linkage systems, as well as group insurance and other relevant schemes. In the first system [26] comparisons were made during and following treatment with benoxaprofen, fenbufen, zomepirac, piroxicam and osmosin (indomethacin). No noteworthy differences were apparent in the proportion of serious complications (i.e. bleeding and perforation) either between the five drugs or between the treatment and follow-up period with any of the five drugs. However, the methods used were insufficiently sensitive to detect the effects the authors sought to observe. Insufficient power may also account for the other studies in which less significant or non-significant associations were reported [27–29].

It should be stressed that so far only relative estimates of the risk of GI bleeding

46

in NSAID users have been measured. Comparable estimates of the excess incidence of bleeding among users of different NSAIDs are not yet available. Only such measures will enable evaluation of whether the increased risk outweighs the expected benefits (principally symptomatic control of usually non-fatal diseases). The need to estimate excess risk has not been sufficiently recognized in the past. A modest elevation in the relative risk can, if the event under study is common, translate into a strikingly high excess risk. The need for epidemiological studies was clearly demonstrated when suggestions were raised (subsequently refuted) that piroxicam has an unusually high risk for causing GI bleeding [30,31].

Analytical Epidemiological Studies of Hematological Reactions to NSAIDs

Most of the available data on hematological NSAID reactions consist of case reports, usually either poorly documented statistics concerning frequencies of spontaneous reports to national registries or hospital surveys of diagnoses such as agranulocytosis, aplastic anemia, or thrombocytopenia. There have been a few attempts at epidemiological studies, including studies of phenylbutazone and oxyphenbutazone in relation to aplastic anemia, and the International Agranulocytosis and Aplastic Anemia Study (IAAAS).

A. Butazones and aplastic anemia
Two estimates of the risk of aplastic anemia due to butazones appeared in the 1970s, one from Sweden by Bottiger and Westerholm [32] and the other from the U. K. by Inman [33]. The first was based on reports to the Swedish Adverse Drug Reactions Committee concerning cases of aplastic anemia thought to be caused by butazones. Denominators were derived from data on Swedish drug sales. The authors calculated a rate of 1 in 99,000, and further estimated that since only one in three cases was actually reported to the Committee, the true rate should be 1 in 33,000. Inman's study was based on data from death certificates and a survey of general practitioners' prescriptions. He estimated that the mortality from aplastic anemia was 2.2 per 100,000 users for phenylbutazone and 3.8 per 100,000 users for oxyphenbutazone.

In both studies, diagnostic criteria were not rigorous, and there were multiple potential sources for bias relating to the ascertainment of the cases and to the information about drug use. In computing the estimates multiple assumptions were made, and it is unclear exactly what the rates referred to, i.e., whether they were per prescription or per exposed individual, nor was there reference to duration of exposure. Therefore it still remains unclear how to interpret the risk described in these reports.

B. The International Agranulocytosis and Aplastic Anemia Study (IAAAS) and NSAIDs

The methods used in this population-based case-control study have been published [34,35]. Drug use, including NSAID use, during the week before onset of clinical illness was compared between 221 confirmed cases of agranulocytosis and 1,425 hospital controls identified by study centers in Jerusalem, Berlin, Ulm, Milan, Barcelona, Sofia, Budapest, and Stockholm. The study base comprised the total experience in these areas in the years 1980–1984, amounting to 80 million person-years. NSAID use from 29 to 180 days before hospital admission was also compared between 113 cases of aplastic anemia and 1,724 controls. For agranulocytosis and aplastic anemia the IAAAS estimated a somewhat elevated rate ratio for salicylates, but these results should be considered tentative and some elaboration of the data should be published shortly. For butazones rate ratio estimates were 3.8 (95% c.l., 1.3–10.7) for agranulocytosis and 8.7 (3.4–22) for aplastic anemia. The excess risk for agranulocytosis was 0.2 cases per million for any "butazone" exposure during a one-week period. For aplastic anemia the estimate was 6.6 cases per million for any exposure during a five-month period.

The estimated rate ratio for indomethacin was 8.9 (c.l., 2.9–28) for agranulocytosis and 12.7 (c.l., 4.2–38) for aplastic anemia, respectively. Excess risks were 0.6 per million for any exposure during a one-week period and 10.1 per million for any exposure during a five-month period. The IAAAS had limited data on the new NSAIDs; however, there was sufficient information to evaluate diclofenac. For that drug there is an association with aplastic anemia (estimated rate ratio, 8.8; c.l., 2.8–27). The excess risk for any exposure during a five-month period was estimated at 6.8 per million.

Conclusions

Limited progress was made during the last decade in pharmacoepidemiological research of adverse reactions to NSAIDs. The most striking discovery was the drug etiology of Reye's syndrome [36]. Estimates of excess risk for agranulocytosis and aplastic anemia from several NSAIDs became available, but these values require confirmation; the ADRs are rare and the excess risks are small. For the *less* rare event of major GI bleeding an association with NSAIDs has been repeatedly shown and rate ratios determined, while estimates of the excess risks are still pending. Aspirin-induced GI bleeding appears to be more common than previously thought, and may also occur after occasional use. Estimates of risk are not yet available either for most NSAIDs or for many ADRs.

48

For any given drug, the evaluation is seldom complete when only one outcome has been studied; moreover, judgments can only be made if comparable data is available for alternative drugs. If one drug is to be replaced by another, the relative risks must be known. The availability of comparable quantitative data for all major adverse events to NSAIDs and the identification of subpopulations carrying higher risks should be goals for the coming decade. A change from pharmacopolitics to pharmacoepidemiology is a formidable challenge for the pharmaceutical industry, health authorities, and academia.

Acknowledgement

The author thanks the Adolfo and Evelyn Blum Fund for Arthritis Research, and Dr. Tanya Livshits and Mrs. Pauline Ziderman, for their valuable help.

References

1. Slone, D., Shapiro, S., and Miettinen, 0.S., Case-control surveillance of serious illnesses attributable to ambulatory drug use. In: Colombo, F., Shapiro, S., Slone, D., and Tognoni, G., (Eds.), *Epidemiological evaluation of drugs*. PSG Publishing Co., Littleton, MA 1977, p. 59.
2. Inman, W.H., Recorded release. In: Gross, F. H. and Inman, W. H., (Eds.), *Drug monitoring*. Academic Press, London 1979.
3. Royal College of General Practitioners' Oral Contraceptive Study, Further analyses of mortality on oral contraceptive users, 1,541 (1981).
4. Giercksky, K., Huseby, G., and Rugstad, H. E., Epidemiology of NSAID-related gastrointestinal side-effects. Scand. J. Gastroenterol **163**, suppl.: 3 (1989).
5. Cole, P., The evolving case-control study. J. Chr. Dis. **32**: 15 (1979).
6. Schlesselman, J.J., Case-control studies: design, conduct, analysis. Oxford University Press, New York 1982, p. 40.
7. MacMahon, B. and Pugh, T.F., *Epidemiology: principles and methods*. Little, Brown, Boston 1970.
8. Fowler, P.D., Aspirin, paracetamol and non-steroidal anti-inflammatory drugs: a comparative review of side effects. Med. Toxicol. **2**: 338 (1987).
9. Hawkey, C.J., Non-steroidal anti-inflammatory drugs and peptic ulcers. Br. Med. J. **300**: 278 (1990).
10. Levy, M. and Heyman, A., Hematological adverse effects of analgesic anti-inflammatory drugs. Hematol. Rev., **4**: 177 (1990).
11. Levy, M., Epidemiological evaluation of rare side-effects of mild analgesics. Br. J. Clin. Pharmacol. **10**: 395s (1980).
12. Douthwaite, A.H., Some recent advancees in medical diagnosis and treatment. Br. Med. J. **1**: 1143 (1938).
13. Levy, M., The epidemiological evaluation of major upper gastrointestinal bleeding in relation to aspirin use. In: Kewitz, H., Roots, I., and Voigt, K., (Eds.), *The epidemiological concepts of clinical pharmacology*. Springer Verlag, Berlin 1987, p. 100.

14. Levy, M., Aspirin use in patients with major upper gastrointestinal bleeding and peptic ulcer disease. N. Engl. J. Med. **290**: 1158 (1974).

15. Coggon, D., Langman, M.J.S., and Spiegelhalter, D., Aspirin, paracetamol and haematemesis and melena. Gut **23**: 340 (1982).

16. Levy, M., Miller, D.R., Kaufman, D.W. et al., Major upper gastrointestinal bleeding. Relation to the use of aspirin and other non-narcotic analgesics. Arch. Intern. Med. **148**: 281 (1988).

17. Kaufman, D.W., personal communication (1990).

18. Faulkner, G., Prichard, P., Somerville, K., and Langman, M.J.S., Aspirin and bleeding peptic ulcers in the elderly. Br. Med. J. **297**: 1311 (1988).

19. Aspirin Myocardial Infarction Study Research Group, A randomized, controlled trial of aspirin in persons recovered from myocardial infarction. J. Am. Med. Assn. **243**: 661 (1980).

20. Persantine-Aspirin Reinfarction Study Research Group, Persantine and aspirin in coronary heart disease. Circulation **62**: 449 (1980).

21. Steering Committee of the Physicians' Health Study Research Group, Final report on the aspirin component of the ongoing Physicians' Health Study. N. Engl. J. Med. **321**: 129 (1989).

22. Sommerville, K., Faulkner, G., and Langman, M.J.S., Nonsteroidal anti-inflammatory drugs and bleeding peptic ulcer. Lancet, i, 462 (1986).

23. Griffin, M.R., Ray, W.A., and Schaffner, W., Non-steroidal anti-inflammatory drug use and death from peptic ulcer in elderly persons. Ann. Intern. Med. **109**: 359 (1988).

24. Smedley, F.H., Traube, M., Leach, R., and Wastell, C., Nonsteroidal anti-inflammatory drugs: retrospective study of 272 bleeding or perforated peptic ulcers. Postgrad. Med. J. **65**: 892 (1989).

25. Langman, M.J.S., Morgan, L., and Worrall, A., Use of anti-inflammatory drugs by patients admitted with small and large bowel perforation and haemorrhage. Br. Med. J. **290**: 347 (1985).

26. Inman W.H., Comprehensive study of five non-steroidal anti-inflammatory drugs. Prescript. Event Monitor. News **3**, 3 (1985).

27. Carson, J.I., Strom, B.L., Soper, K.A., West, S.I., and Morse, M.I., The association of non-steroidal anti-inflammatory drugs with upper gastrointestinal tract bleeding. Arch. Intern. Med. **147**: 85 (1987).

28. Beard, K., Walker, A.M., Perera D.R., and Jick, H., Nonsteroidal anti-inflammatory drugs and hospitalization for gastroesophageal bleeding in the elderly. Arch. Intern. Med. **147**: 1621 (1987).

29. Beardon, P.H.G., Brown, S.V., and McDevitt, D.G., Gastro-intestinal events in patients prescribed non-steroidal anti-inflammatory drugs: A controlled study using record linkage in Tayside. Q. J. Med. **71**: 497 (1989).

30. Bortnichak, E.A. and Sachs, R.M., Piroxicam in recent epidemiologic studies. Am. J. Med. **81**: 44 (1986).

31. Rossi, A.C., Hsu, J.P., and Faich, G.A., Ulcerogenicity of piroxicam. An analysis of spontaneously reported data. Br. Med. J. **294**: 147 (1987).

32. Bottiger, L.E. and Westerholm, B., Blood-induced blood dyscrasias in Sweden. Br. Med. J. **3**: 339 (1973).

33. Inman, W.H., Study of fatal bone marrow depression with special reference to phenylbutazone and oxyphenbutazone. Br. Med. J. **1**, 1500 (1977).

34. International Agranulocytosis and Aplastic Anemia Study, The design of a study of the drug etiology of agranulocytosis and aplastic anemia. Eur. J. Clin. Pharmacol. **24**: 833 (1983).

35. International Agranulocytosis and Aplastic Anemia Study, Risks of agranulocytosis and aplastic anemia. A first report of their relation to drug use with special reference to analgesics. J. Am. Med. Assn. **256**: 1749 (1986).
36. Starko, K. M., Ray, C. G., and Dominguez, L. B., Reye's syndrome and salicylate use. Pediatrics **66**: 859 (1980).

Adverse Reactions to NSAID: Clinical Pharmacoepidemiology
ed. by M. Kurowski
© 1992 Birkhäuser Verlag Basel

The SPALA Project – An Intensive Monitoring System for Non-Steroidal Anti-Inflammatory Drugs

The SPALA Project Team
Presentation by M. Kurowski

Verein zur Langzeituntersuchung von Arzneimittelwirkungen auf dem Gebiet der Rheumatologie e. V., c/o Institut für Pharmakologie und Toxikologie der Universität Erlangen-Nürnberg, Universitätsstraße 22, D-W–8520 Erlangen, Germany

Introduction

The SPALA ("Safety Profile of Antirheumatics in Long-Term Administration") project, planned and initiated in 1987–1988 by members of F. Hoffmann-La Roche AG (Basel, Switzerland) and the Institut für Pharmakologie und Toxikologie der Universität Erlangen-Nürnberg (Germany), was intended to identify, collect, classify, and quantify adverse events (AEs) occurring during or after treatment with non-steroidal anti-inflammatory drugs (NSAIDs). Data were collected from a cohort of approximately 30,000 patients who received NSAIDs during hospitalization or ambulatory treatment in participating rheumatological centers distributed over a wide geographical area in Germany, Austria, and the German-speaking part of Switzerland (Tab. 1). The project was conducted in conformity with the applicable laws in each of these countries. Detailed descriptions of SPALA have already been published [1,2]. Briefly, data were collected by trained physicians and nurses specifically employed for this project. A three-part questionnaire was utilized to document patients' medical histories, treatments during the observation period, and AEs. The latter were registered once a week after a simple standard question to the hospitalized patients. Additional information was obtained from physicians, nurses, and medical records. Outpatients were asked the same question during every visit to the medical centers. Questionnaires were mailed to a professional data processing company (Post-Marketing Surveillance (PMS), Ltd., London, a subsidiary of IMS International); data were entered into a computer, classified, and analyzed. Classification of diseases (according to the World Health Organization International Classification of Diseases (WHO

Table 1. Persons and institutions participating in the SPALA project.

Project plan R. Lanz, H. Fenner (Basel); M. Kurowski, K. Brune (Erlangen).
Project management M. Kurowski (Erlangen).
Advisory board E. Weber (Heidelberg); K. Brune (Erlangen); M. Franke (Baden-Baden); H. Kewitz (Berlin); K. H. Kimbel (Köln); R. Repges (Aachen).
"In case" experts M. Schneider, Gastroenterology (Erlangen); T. Ruzicka, Dermatology (München); J. Mann, Nephrology (Nürnberg); W. Heit, Hematology (Ulm).
Participating centers and physicians in charge of monitoring *Germany* U. Botzenhardt (Bremen); D. Jentsch, E. Keck, K. Miehlke (Wiesbaden); G. Josenhans (Bad Bramstedt); E.-M. Lemmel (Baden-Baden); H. Menninger (Bad Abbach); M. Schattenkirchner (München); H. Sörensen (Berlin); T. Stratz (Bad Säckingen); H. Zeidler (Hannover).
Austria R. Eberl, A. Dunky (Wien); G. Kolarz (Baden); O. Scherak (Baden); N. Thumb (Baden).
Switzerland P. Mennet (Rheinfelden); W. Müller (Basel); F. J. Wagenhäuser, M. Felder (Zürich).
Data management G. Kiep, K. Szendey (Frankfurt).

ICD), 9th revision) and AEs (according to the WHO Adverse Reaction Terminology (ART)) was performed by PMS, Ltd. in collaboration with the project management. Immediately reportable adverse events (IRAEs) were reported directly to the responsible authorities, the manufacturers of the administered drugs, and the head of the SPALA advisory board. The planning, execution, and evaluation of SPALA was supervised by an advisory board consisting of experienced specialists in relevant disciplines.

Experts in gastroenterology, hematology, dermatology, and nephrology were available to assist in the diagnosis of severe adverse drug reactions (ADRs). Participating persons and institutions are listed in Table 1. This report contains a summary of the collected data; further analyses of these data will subsequently be performed and published.

Definitions

Adverse event (AE): every clinically relevant deterioration in a patient's condition, whether or not it appears to be connected with the medication; every symptomologic deterioration and every appearance of irregularities in the clinical picture or in laboratory values considered to be clinically relevant by the doctor and temporally (during or following), but not necessarily causally, related to the medication.

Immediately reportable adverse event (IRAE): every adverse event which
– is fatal;
– is life-threatening or permanently disabling; or
– requires hospitalization or prolongation of hospitalization.

Adverse drug reaction (ADR): every detrimental and unintended reaction to a drug administered at a dosage recommended for prophylaxis, diagnosis, or therapy (*excluding* the absence of any therapeutic effect).

Observation period: period of NSAID treatment of a patient at one of the monitoring centers. For hospitalized patients, the observation period ended no later than the day of discharge from the clinic. For ambulatory patients the observation period ended no later than the date of the *last* consultation at the clinic, defined as the visit followed by a period 6 weeks during which the patient did not seek an additional consultation. Spontaneous reports of AEs were recorded up to 6 weeks after the end of the observation period, though such reports were no longer collected systematically.

Treatment case (TC): under certain conditions patients could be included and registered more than once, either by the start of a new NSAID therapy after the end of an observation period, by patient transfer from one participating hospital to another, or by a change in patient status from ambulatory to in-hospital or vice versa. The primary data evaluation and results refer to TCs. Subsequent analysis will include the reduction of TCs to patients.

Adverse event case (AEC): a TC who experienced at least one AE is denoted as an AEC. The number of AEs (9,480) was greater than the number of AECs (5,457), with an overall AE/AEC ratio of 1.7.

Data sources

The following sections describe the participating centers, the patients, the diseases, the prescriptions, and the AEs. Table 2 presents an overview of the database for completely documented cases.

Description of the centers

The medical centers participating in SPALA were located in Austria (4), Germany (9), and Switzerland (3). The number of monitored TCs per center ranged from 506 to 4,482. Table 3 illustrates some differences between the various centers regarding their contributions to the final results. The differences in AE/TC ratios (last column of Table 3) at the participating centers require some further explanations, as consideration of center effects in the AE reporting frequency (AE/TC) is a prerequisite for an evaluation of the summarized data. Possible explanations for these differences include different patient populations, different diseases, different prescription patterns for NSAIDs, and different intensities of and variations within AE monitoring. The observed AE rates for hospitalized patients, approximately double that for outpatients, might be due to the more intensive monitoring and the collection of more comprehensive medical information during hospitalization.

Table 2. Overall Results.

	Male	Female	Total
No. of TCs (at least one NSAID prescription)	10,504	18,560	29,064
No. and percentage of TCs with at least one AE (AECs)	1,439 (13.7%)	4,018 (18.1%)	5,457 (16.5%)
No. of diagnoses indicating NSAID treatment	11,050	19,507	30,557
No. of NSAID prescriptions	12,946	23,201	36,147
No. of AEs	2,257	7,223	9,480
Average no. of NSAID prescriptions per TC	1.2	1.3	1.2
Average no. of AEs per AEC	1.6	1.8	1.7
Average no. of NSAID prescriptions per AE	5.7	3.2	3.8

Table 3. Breakdown of participating centers in terms of TCs, NSAID prescriptions, and AEs.

	Treatment cases (TC)	NSAID prescriptions		Adverse events	
	#	#	#/TC	#	#/TC
Baden (2 centers)	1,061	1,447	1.36	439	0.41
Baden (1 center)	506	795	1.57	121	0.24
Bad Abbach	1,489	1,821	1.22	439	0.29
Bad Bramstedt	4,482	5,668	1.26	1,964	0.44
Bad Säckingen	547	754	1.38	154	0.28
Baden-Baden	3,569	4,422	1.24	1,159	0.32
Basel	2,115	2,909	1.38	569	0.27
Berlin	1,707	2,225	1.30	1,714	1.00
Bremen	1,495	1,930	1.29	782	0.52
Hannover	1,211	1,358	1.12	190	0.16
München	1,674	1,907	1.14	452	0.27
Rheinfelden	938	1,079	1.15	152	0.16
Wien	1,195	1,566	1.31	289	0.24
Wiesbaden	2,597	2,954	1.14	529	0.20
Zürich	4,478	5,312	1.19	527	0.12
Out-hospital	9,095	10,828	1.19	1,798	0.20
In-hospital	19,969	25,319	1.27	7,682	0.38
Total	29,064	36,147	1.24	9,480	0.33

Description of the patients

Data have not been assigned to specific patients; rather, each datum is associated with a particular TC. Table 4 illustrates the TC distribution by sex, country, and hospitalization status. The gender distribution in the three countries varies considerably, the highest portion of female cases being in Austria and the lowest in Switzerland. A somewhat higher proportion of female cases was registered among in-hospital TCs than among ambulatory TCs.

The age distribution for TCs and AECs according to gender is presented in Table 5. The age distribution of TCs differs from that of AECs, with a shift towards

56

Table 4. TCs by gender, country, and hospitalization status.

		Male	Female
Country	Austria	854	1,908
	Germany	6,152	12,619
	Switzerland	3,498	4,033
Status	In-hospital	6,645	13,324
	Out-hospital	3,859	5,236
Totals		10,504	18,560

Adverse event cases
Age distribution

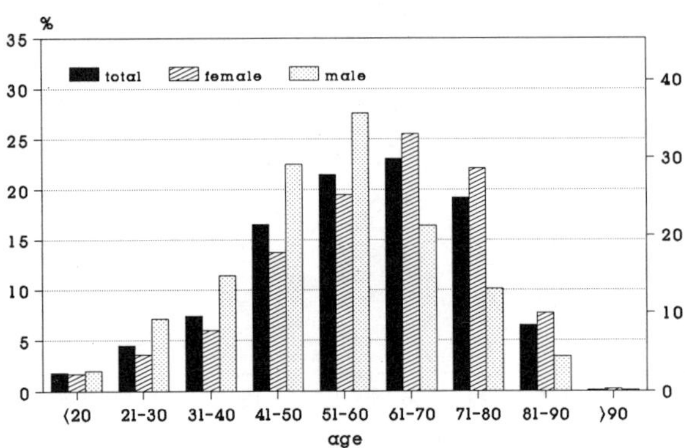

Treatment cases
Age distribution

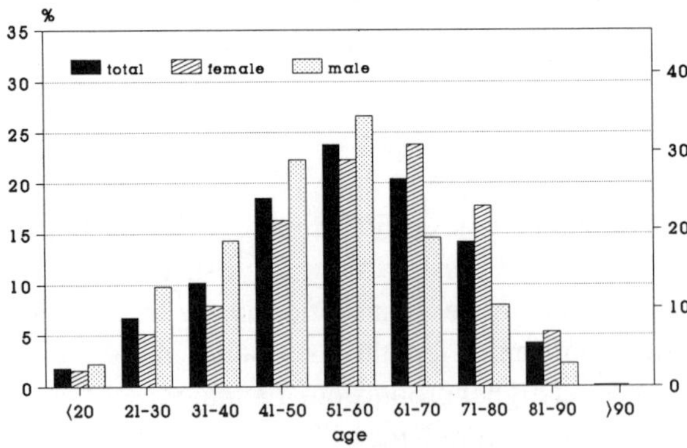

Fig. 1.

Table 5. Percentage of different age classes, separated by gender, for TCs and AECs.

Age class (years)	< 20	21–30	31–40	41–50	51–60	61–70	71–80	81–90	> 90
TCs									
Male (%) Mean age: 50 years	2.2	9.8	14.3	22.3	26.6	14.6	8.0	2.2	< 0.1
Female (%) Mean age: 58 years	1.6	5.1	7.9	16.3	22.3	23.8	17.7	5.3	0.1
Total (%) Mean age: 55 years	1.8	6.8	10.2	18.5	23.8	20.4	14.2	4.2	0.1
TCs with at least 1 AEC									
Male (%) Mean age: 53 years	2.0	7.1	11.4	22.5	27.0	16.4	10.1	3.4	0.1
Female (%) Mean age: 61 years	1.7	3.6	6.0	13.7	19.5	25.5	22.1	7.7	0.2
Total (%) Mean age: 59 years	1.8	4.5	7.4	16.0	21.5	23.1	19.0	6.5	0.1

higher ages in the latter group (Fig. 1). For both sexes the average age difference between TCs and AECs was 3 years.

Diseases

Among rheumatic diseases, degenerative disorders generally prevail. However, the spectrum of diseases treated at medical centers participating in SPALA was clearly different from what is more commonly observed in the population at large. The diagnoses of the rheumatic disorders have been registered and classified according to the WHO ICD code. A large variety of diseases is covered by the term "rheumatic disorders"; therefore, to give an overview of the diagnoses which led to treatment with NSAIDs, a rough breakdown into four classes was used:
(i)   primarily inflammatory rheumatic disorders;
(ii) primarily degenerative rheumatic disorders;
(iii) other rheumatic disorders;
(iv) other disorders.

The latter group is comprised of disorders which are not strictly classified as "rheumatic", but for which an NSAID was indicated. The distribution of the total 30,557 diagnoses is depicted in Figure 2.

Indications for NSAID-treatment
(total n=30557 diagnoses)

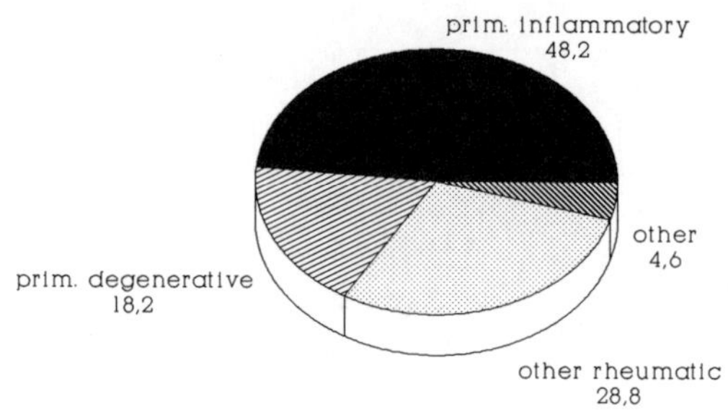

Fig. 2.

Treatment with NSAIDs

The number and variety of available NSAIDs was substantially different in the three participating countries. The prescribed preparations, including generic products, numbered 35 in Austria, 76 in Switzerland, and 182 in Germany. In accordance with German law, the only criterion for the use of a drug in observational PMS studies like SPALA is medical necessity, and the physician's selection of medication must not be otherwise influenced. Thus the proportions of prescribed medications reflect the actual prescription patterns in the participating centers. Despite the variety of drugs available, the four most commonly-prescribed NSAIDs accounted for > 70% of the total of 36,147 prescriptions (Fig. 3).

Table 6 lists the ten most frequently prescribed drugs, which amounted to 32,937 (91%) out of a total number of 36,147 prescriptions. These prescriptions include all available formulations and strengths. If, during the course of treatment, it was found that a patient never used the prescribed NSAIDs due to bad compliance or any other reason, that patient was excluded from the evaluation.

Patients were usually prescribed new drugs when commencing in- or out-hospital treatment. The average duration of treatment was relatively short; therefore, long-term therapy effects could only be monitored in a small population. The percentage distribution of NSAID prescriptions with regard to the length of treatment was:

1 day	< 1week	< 3weeks	< 6weeks	≥ 6 weeks
13.9%	14.3%	30.0%	32.3%	9.5%

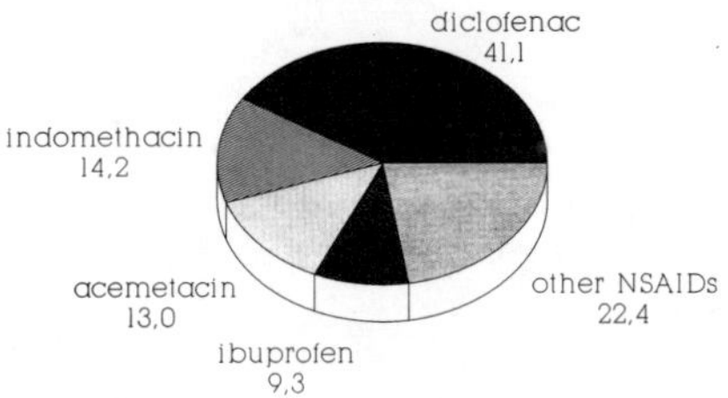

The total number of AE is 9480; some AE
refer to more than one NSAID prescription

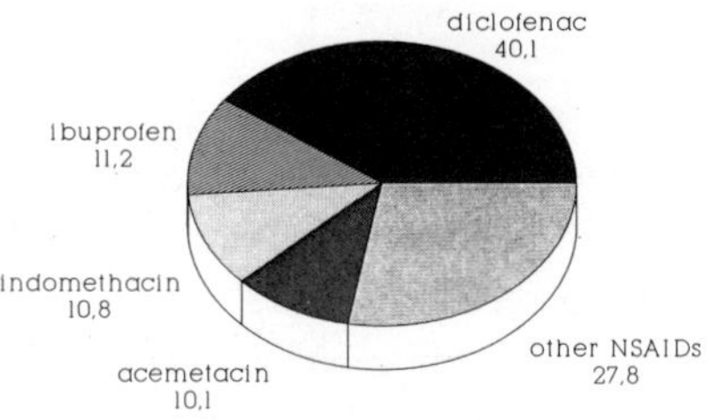

Fig. 3.

Each prescription was registered at the beginning and end of treatment, including the reason for discontinuation. In addition, routes of administration as well as dosage strengths and schedules were recorded, but this first evaluational report does not take the doses into account.

Adverse events

The physicians responsible for data collection in each center were trained continuously in the monitoring and documentation of AEs in order to keep similar standards in the various centers. Despite these efforts, differences in AE recognition and reporting cannot be excluded.

Table 6. NSAID prescriptions.

Drug	No. of prescriptions
Diclofenac	14,477
Ibuprofen	4,037
Indomethacin	3,896
Acemetacin	3,633
Piroxicam	1,645
Acetylsalicylic acid	1,211
Ketoprofen	1,183
Tenoxicam	1,075
Naproxen	1,067
Etofenamate	713
Subtotal:	32,937
Total no. of prescriptions	36,147

A total of 9,480 AEs were reported. Some of these refer to more than one NSAID; if one NSAID was taken up to 6 weeks before the onset of an AE, *all* of the administered NSAIDs were defined as linked to the event. The events were classified according to the WHO ART under the guidance of an experienced supervisor. Table 7 lists 10 NSAIDs associated with AEs and the number of reported AEs per drug. The proportions of total prescriptions and of AEs for the four most frequently-prescribed NSAIDs are compared in Figure 3, from which

Table 7. AE distribution of the 10 NSAIDs most frequently associated with AEs.

Drug	No. of prescriptions associated with AEs
Diclofenac	4,891
Indomethacin	1,693
Acemetacin	1,553
Ibuprofen	1,110
Piroxicam	488
Ketoprofen	448
Tenoxicam	359
Naproxen	282
Acetylsalicylic acid	250
Pirprofen	188

it will be seen that there is a good correlation between the frequency of NSAID prescriptions and the frequency of AE reports.

The classification of AEs according to system-organ classes revealed additional characteristics exhibited by the group consisting of diclofenac, indomethacin, acemetacin, and ibuprofen. Despite the similarity of AE patterns for all NSAIDs, some drug-specific differences were noted (Figs. 4 and 5), e.g. central nervous system AEs accompanying medication with indomethacin. Also, the large percentage of gastrointestinal (GI) tract AEs observed with acemetacin remains to be explained (Fig. 4).

Generally speaking, the GI tract is by far the most frequently-affected system, followed by the skin, the nervous system, the body as a whole, and the respiratory

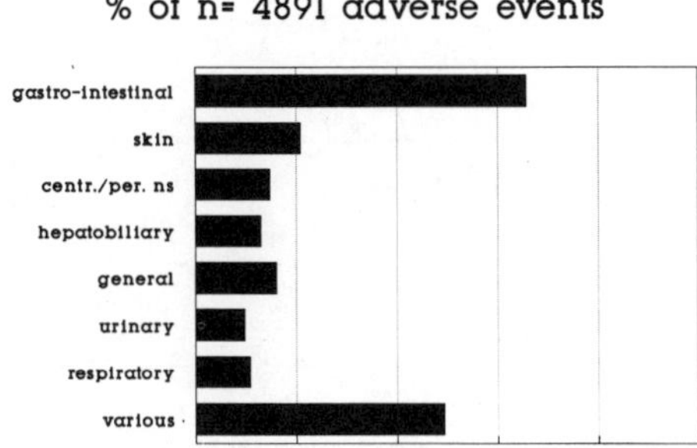

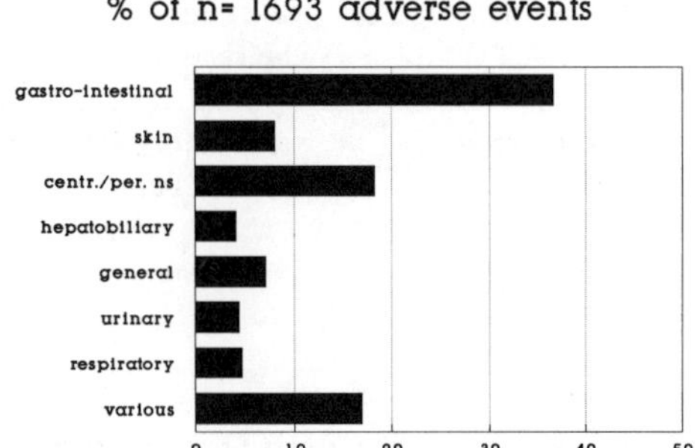

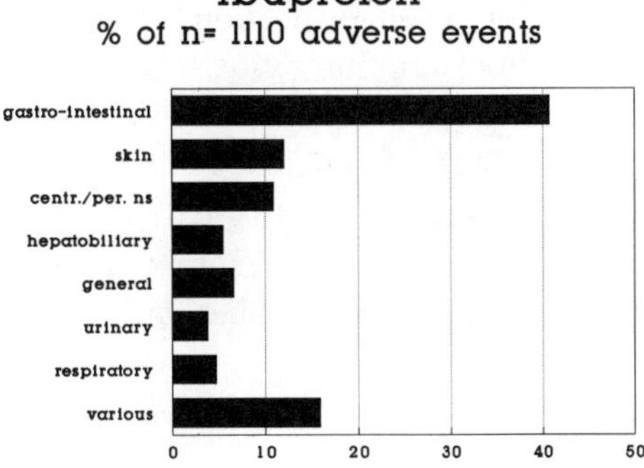

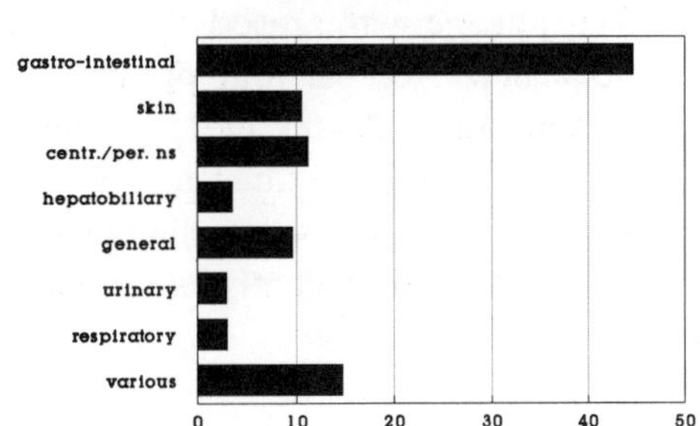

Fig. 4.

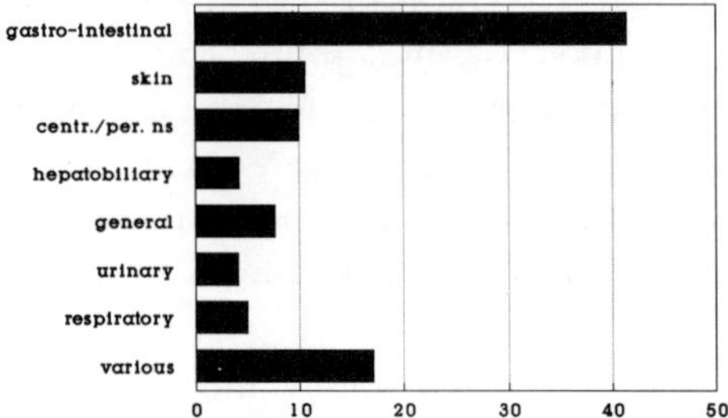

Fig. 5.

Abbreviation	WHO system-organ class
gastrointestinal	gastrointestinal system disorders
skin	skin and appendages disorders
CNS/PNS	central and peripheral nervous system disorders
hepatobiliary	liver and biliary system disorders
general	body as a whole – general disorders
urinary	urinary system disorders
respiratory	respiratory system disorders
various	AEs affecting systems not included in the above

system. Since the WHO category "body as a whole – general disorders" comprises all forms of edema, it is conceivable that such effects caused by NSAIDs via the kidneys and the cardiovascular system fell into this system-organ class.

At the present stage of analysis, the final step in the quantification of AEs is their linkage with prescriptions, expressed as AE/prescription ratios. These AE frequencies for the four principal NSAIDs and the most commonly-affected system-organ classes are presented in Table 8. These frequencies should be regarded as rough estimates of risks, to be used and interpreted with great caution, as many possible confounding factors have not yet been critically examined, and should thus serve merely as a guideline for further studies and evaluations.

Immediately reportable adverse events

In order to meet the various national legal requirements for reporting serious adverse drug reactions and to have a means of quickly disseminating information

Table 8. Frequencies of AEs (AE/NSAID prescriptions).

System-organ class	Diclofenac	Ibuprofen	Indomethacin	Acemetacin
Gastro-Intestinal system	14.1%	11.2%	15.9%	19.1%
skin and appendages	3.5%	3.3%	3.5%	4.5%
Central and peripheral nervous system	2.5%	3.0%	7.9%	4.8%
Liver and biliary system	2.2%	0.7%	1.8%	1,5%
Body as a whole – general	2.7%	2.2%	3.1%	4.1%
No. of NSAID prescriptions	14,477 (100%)	4,037 (100%)	3,896 (100%)	3,633 (100%)

about serious AEs, the participating medical centers reported all such events directly to the project management. All deaths, life-threatening or permanently-disabling events, or situations necessitating or prolonging hospitalization according to the project guideline had to be reported within one working day. The project manager, in turn, informed the authorities, the manufacturers, the advisory board, and other concerned institutions.

A total of 220 IRAEs were reported in the course of this project (2.3% of all registered AEs or 0.8 IRAE/100 TCs). Thirty-four deaths were reported, but none of these are believed associated with NSAID consumption. Fifty-six IRAEs were considered to be associated with the use of NSAIDs. Table 9 shows the distribution of IRAEs among the prescribed NSAIDs and the WHO system-organ classes. Some patients received more than one NSAID, either consecutively or simultaneously, before onset of an IRAE, thus the higher number of prescriptions (71) compared to the number of IRAEs.

Conclusions

SPALA was designed for the health systems and the legal situations in those countries where the participating hospitals were located. As an instrument for recognizing and quantifying AEs it produced results similar to the spontaneous reporting system of the Drug Commission of the German Medical Profession (Arzneimittelkommission der deutschen Ärzteschaft). In recent publications [3,4] it was noted that both SPALA and the German Spontaneous Reporting System produced similar results. Experience with the installation and operation of such systems, especially in Germany, is limited considering the size of the NSAID market and the manufacturing potential for these drugs; this discrepancy can be

Table 9. Association of 56 IRAEs with prescribed NSAIDs and WHO system-organ classes.

Prescribed NSAIDs	No. of prescriptions with IRAEs	System-organ class affected	No. of IRAEs
Diclofenac	21	GI system	21
Ibuprofen	11	Blood (red cells, white cells, platelets)	18
Indomethacin	7		
Acemetacin	5	Urinary system	5
Piroxicam	5	Respiratory system	5
Tenoxicam	5	Body as a whole	2
Acetylsalicylic acid	5	CNS and PNS	1
Ketoprofen	3	Cardiovascular system	1
Pirprofen	3	Skin and appendages	1
Naproxen	2	Musculoskeletal system	1
Tiaprofenic acid	2	Liver and biliary system	1
Benorylate	1		
Diflunisal	1		

explained in terms of costs, legal obstacles, and the complicated nature of the health system.

Some major problems with SPALA were a consequence of the legal conditions imposed on participating hospitals by their host countries. For instance, no control group without NSAID treatment was observed simultaneously to allow for comparison of AE rates in patients exposed to different NSAIDs while precluding a comparison of incidence rates of AEs in the exposed group with those in an unexposed group, i.e. the calculation of the relative risks of NSAID treatment. The data base was different for each drug, since the numbers of prescriptions varied considerably. Also, some AEs were not recorded as a consequence of the transfer of patients to other wards or hospitals.

The distribution of diseases among patients was certainly not representative of NSAID consumers in general. For instance, there was a high proportion of primarily inflammatory rheumatic diseases due to the inclusion of highly-specialized centers of internal medicine in this study. This distribution would have been considerably different had orthopedic departments also been included in SPALA. Observation of AEs by practitioners in private medical practice was not included in this project; thus, some of the results may not be applicable to NSAID use under these conditions.

A cursory examination of the data reveals different kinds of clusters, which have yet to be analyzed. For instance, preferences for certain medications or increased alertness for certain AEs in some centers could skew the AE profiles. Some of these distorting factors will be analyzed and accounted for in the course of subsequent data assessment.

During the period of data collection it was found that differences between AE profiles for different NSAIDs tended to diminish with increasing numbers of prescriptions. No new types of AEs or strikingly different patterns were recognized. The value of this project can be seen in the complete documentation of the safety of NSAIDs for a large cohort of patients. This type of study became necessary after some NSAIDs had to be withdrawn from the market over the past few years due to ADRs. It is hoped that the results of SPALA will help to reestablish confidence in the safety of these important drugs, which chronically ill patients need to cope with their diseases. Such studies result in both physicians and patients obtaining more detailed information about the type and frequency of AEs associated with drug intake. In addition, the SPALA database contains over 200 stored items for each TC, which can be combined in different ways in order to answer particular questions concerning drug risks.

Acknowledgement

SPALA was generously funded by F. Hoffmann-La Roche AG (Basel).

References

1. The Design of SPALA (Safety Profile of Antirheumatics in Long- Term Administration). Eur. J. Clin. Pharmacol. **34**: 529 (1988).
2. Sicherheitsprofil von Antirheumatika bei Langzeitanwendung (SPALA). Münch. med. Wochenschr. **34**: 103 (1989).
3. SPALA–Sicherheitsprofil von Antirheumatika bei Langzeitanwendung. Dtsch. Ärztebl. **87**: 37 (1990).
4. Lasek, R., Mathias, B., and Tiaden, J.D., Erfassung unerwünschter Arzneimittelwirkungen. Dtsch. Ärztebl. **88**: 35 (1991).

Adverse Reactions to NSAID: Clinical Pharmacoepidemiology
ed. by M. Kurowski
© 1992 Birkhäuser Verlag Basel

Toxicity Analysis of Mild Analgesics in Clinical Use

E. Weber

*Department of Clinical Pharmacology, University of Heidelberg,
Bergheimer Straße 58, D-W-6900 Heidelberg, Germany*

In 1971 a research group in the Department of Clinical Pharmacology at the University of Heidelberg (Germany) started to assess phenomena which the attending physicians suspected to be adverse drug reactions (ADRs) in the University Hospital. Different methods of documentation and evaluation were checked. Over the entire period of the project (1971–1989) a clinical pharmacologist and a specially-trained staff member collected reports on ADRs observed by doctors during their twice-weekly ward rounds. At the end of 1989 the project had to be abandoned due to lack of funds.

Between 1971 and 1979 about 63,000 hospital admissions were surveyed; from 1980 through 1987 more than 77,000 admittances (corresponding to about 60,000 patients) were analyzed. In the latter sample non-narcotic analgesics (NNAs), especially non-steroidal anti-inflammatory drugs (NSAIDs) were *not* prescribed very often: in 1980 one of these drugs had been prescribed at least once in 27% of the hospitalizations, but by 1987 the frequency had dropped to 17%. Very few ADRs were observed with NNAs, and practically none of the ADRs that were observed were classified as "severe".

To give a quantitative idea of the ADRs due to NNAs, reference is made to data covering the period 1971–1980 [1]. Between 1971 and 1978 all patients' medical charts were reviewed and the drugs precribed were recorded. In 1978 a detailed, comprehensive analysis of drug prescribing was undertaken, and for the next two years drug utilization was precisely measured. From the drug utilization data the incidence of ADRs was calculated.

Eleven thousand three hundred ADRs were reported between 1971 and 1980. In 101 patients (54 men, 47 women) ADRs due to propyphenazone and aminophenazone (11), metamizol (dipyrone) (43) and acetylsalicylic acid (ASA, aspirin, used only as an analgesic) (45), or combinations of these drugs (2) were observed. Metamizol, propyphenazone, and ASA had been prescribed during 11,500, 3,000, and 3,500 hospitalizations, respectively. The overall incidence of ADRs for these substances varied between 0.9% and 2.2% during the period under review, except for 1973 when the ADR frequency rose to 4.0%. Three life-threatening ADRs were reported. ASA is believed to have provoked gastrointestinal bleeding (GIB) leading to a hemoglobin value of 6.3 g/100 ml in a 54-year-old woman suffering from polycythemia vera. A 64-year-old woman had to be admitted for emergency surgery due to massive bleeding from a duodenal ulcer after treatment with ASA. Finally, a 42-year-old man who had been treated with metamizol after a heart valve operation suffered sepsis due to agranulocytosis.

When reviewing the above data, it should be borne in mind that the mean value for a hospital stay was 13.9 days in 1975 and 12.1 days in 1978, so that the duration of treatment was relatively short. In 1978 the mean duration of treatment with metamizol was 3 days (range, 2–32), with aminophenazone 3 days (range, 2–54), and with ASA 6 days (range, 2–41). The incidence rate for ADRs followed the same order: 0.3% for metamizol, 0.3% for aminophenazone, and 0.8–1.3% for ASA. Also, the dosages used were very low; mean daily doses for metamizol, aminophenazone, and ASA were 0.820 g, 0.222 g and 1.059 g, respectively.

Between 1971 and 1978 there were 37 hospital admissions due to ADRs after using an NNA. Metamizol was involved in 7 cases (including one case each of agranulocytosis, Lyell's syndrome, shock, and hypotension); 6 ADRs were attributable to propyphenazone (including one case each of panmyelophtisis, pancytopenia, agranolocytosis, anaphylactic shock, and status asthmaticus); and 21 ADRs (including 2 cases of peptic ulcer, 13 cases of GIB, and one case each of pancytopenia, agranulocytosis, thrombocytopenia, and macrohematuria) were associated with ASA. Combinations of these three drugs were involved in additional incidences of ADRs.

Hospital admissions due to GIB in 1988 and 1989

In 1988 and 1989 a special effort was undertaken to document in detail all severe ADRs. Ninety-nine ADRs after treatment with NSAIDs were recorded for 1988 and 120 for 1989. A substantial portion of which (32 in 1988 and 20 in 1989) were due to GIB and required hospitalization. Only 5 GIB ADRs occurred in-hospital, mostly in connection with thrombolytic procedures. These 57 cases (32 + 20 + 5) were analyzed in an effort to answer the following questions:
– At what point did bleeding occur after starting treatment with an NSAID?
– Were peptic ulcers detectable? If so, in what percentage of cases?
The 57 cases were allocated to 3 groups (Tab. 1). Group 1 comprised all cases treated with non-aspirin NSAIDs; other drugs were also used, and in some patients interactions with drugs promoting bleeding are likely to have occurred. Group 2 consisted of patients treated with ASA as well as other drugs; none of these other drugs are known to support GIB. Group 3 consisted of all patients to whom ASA

Table 1. Patients suffering from GIB treated with NSAIDs.

Year	Group 1	Group 2	Group 3	Totals
1988	8	20	3	31
1989	5	14	7	26
Totals:	13	34	10	57

Group 1: patients treated with NSAIDs (except ASA); partly concomitant medication, some of which may promote bleeding.

Group 2: patients treated with ASA; partly concomitant medication, none of which is known to promote bleeding.

Group 3: patients treated with ASA and other drugs, all of them known to promote bleeding.

was co-administered with substances known to enhance GIB. NSAIDs were administered in standard daily doses. In nearly all patients treated with ASA, less than 1 g was taken; in most cases the dosage was 500 mg.

Tables 2–4 show at which moment GIB occurred in relation to the onset of analgesic treatment and in how many patients gastric or duodenal ulcers were demonstrable. The numbers in Groups 1 and 3 suggest a slight tendency for onset of GIB within the first few days of treatment. This is in accordance with the fact that patients in these groups received NSAIDs plus other drugs, interactions between which were likely to promote GIB. The percentage of patients with ulcers in Groups 1, 2, and 3 was high: ulcers were demonstrable in 9/13 patients in Group 1, in 28/34 patients in Group 2, and in 7/10 patients in Group 3, for a mean percentage of 77%. Data concerning the onset of bleeding are summarized in Fig. 1, and strengthen the impression of an early onset of GIB if other bleeding-promoting drugs are involved and a delay in GIB when bleeding-promoting drugs are not co-administered.

The results of this survey are not unassailable, nor are they based on a sufficiently large number of cases to allow an unequivocal, unambiguous statement of the relationship between NSAIDs and GIB. However, as a consequence of these results, it is of interest to evaluate further cases. Indeed, there are other indications that GIB generally does not occur within the first few days of treatment with NSAIDs provided no contraindications like ulcers or drug interactions are present. On this basis, the prophylactic use of ASA as an antithrombotic agent was investigated in a number of large prospective studies about 20 years ago. Analyzing all prospective randomized, placebo-controlled studies in which ASA was tested as an anti-thombotic agent for 2 or 3 weeks after different kinds of surgery in a total of 5,444 patients, none of the few cases of GIB were clearly attributable to ASA. The dosage in most instances was 1.5 g ASA per diem [2]. In his analysis

Table 2. Time interval between the start of treatment with NSAIDs and the onset of GIB in Group 1 patients. For characterization of the Groups, see legend of Table 1.

	Duration of therapy				Total
	7–21 days	Several –6 months	1– several years	Unknown	
No. of Group 1 patients	7	2	3	1	13
No. of Group 1 patients with ulcers	4/7	2/2	3/3	1/1	

Table 3. Time interval between the start of treatment with ASA and the onset of GIB in Group 2 patients. For characterization of the Groups, see legend of Table 1.

	Duration of therapy				Total
	1–24 days	1–10 months	1–11 years	Unknown	
No. of Group 2 patients	9	8	15	2	34
No. of Group 2 patients with ulcers	7/9	7/8	12/15	2/2	

Table 4. Time interval between the start of treatment with ASA and the onset of GIB in Group 3 patients. For characterization of the Groups, see legend of Table 1.

	Duration of therapy				Total
	1–13 days	4 months	1 year	Unknown	
No. of Group 1 patients	4	2	1	3	10
No. of Group 1 patients with ulcers	3/4	1/2	1/1	2/3	

of a study in which ASA was prophylactically used, Bousser et al. [3] concluded that most GIB occurred only after about 11 months. Carson et al. [4], examining computerized Medicaid data, found that in 47,136 patients treated with NSAIDs (but not ASA) hospitalization for GIB (155 patients) peaked after the fourth prescription for an NSAID, or roughly four months after the commencement of treatment. A plot of patient hospitalization vs. prescription number generated a quadratic curve that declined precipitously after the fourth prescription.

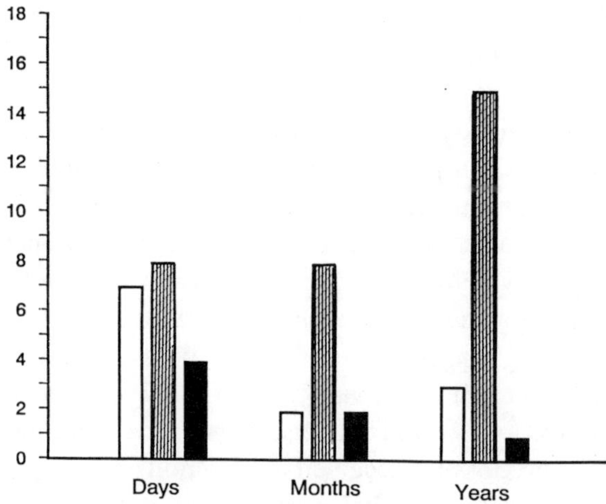

Figure 1. Time interval between the start of treatment with different NSAIDs and the onset of GIB. For characterization of the Groups, see legend to Table 1.

Evidence suggests that GIB due to non-antirheumatic doses of ASA is dose-dependent

Questions about the ASA dose dependence of GIB are not as banal as they might seem. Apart from some correlations between GIB and damaged gastric mucosa, and a certain disposition to GIB arising from alcohol abuse and some diseases of the GI tract, little is known about the factors contributing to GIB or which persons are particularly at risk. Weber et al. [5] reviewed 62 long-term studies on ASA, in which the drug had been used for different purposes, mainly as secondary prophylaxis of myocardial infarction and cerebral ischemia, in daily doses ranging from 900–1,500 mg or 100–325 mg. Nineteen of these 62 studies met the following criteria: prospective, placebo-controlled, double-blind, randomized; one group treated with ASA or ASA + dipyridamole (a coronary vasodilator); adequate documentation of ADRs. Analysis of these 19 long-term studies indicated that an ASA dose dependency for GIB exists, though unambiguous confirmation of this conclusion could not be obtained. In the U.K. TIA study [6], in which daily doses of 300 mg and 1,500 mg of ASA were tested against a placebo, fewer ADRs, including GIB, were found at the lower ASA dose. Since the same clinical effect seemed to be obtained with low doses (100–325 mg) of ASA as with conventional doses (900–1,500 mg), there is no doubt that, at least for purposes of drug safety, low doses should be used when prophylaxis is intended.

Implications of pre-operative administration of ASA for patients undergoing coronary artery bypass grafting

In a prospective randomized placebo-controlled study, either 325 mg of ASA once or three times daily, or 325 mg of ASA once daily + 75 mg of dipyridamole or 267 mg of sulfinpyrazone three times daily, or a placebo were administered. The cohort consisted of 772 pre-operative patients due to undergo coronary artery bypass grafting. Sethi [7] convincingly demonstrated that, compared to the non-ASA groups, the ASA groups had greater volumes in their chest tube drainage, more post-operative bleeding episodes requiring re-operation, and a longer interval between completion of cardiopulmonary bypass and wound closure, meaning that more time was necessary to achieve adequate hemostasis in the operating room. Patients in the ASA groups also received more blood and blood products (except whole blood). The overall in-hospital mortality rate and 30-day operative mortality rate was 2.3%. It was 2.5% for patients in the ASA groups and 2.0% for patients in the non-ASA groups (p = 0.619). These results clearly indicate that practitioners should refrain from prescribing ASA if cardiac surgery is to be performed within the next few days; moreover, it is a general rule that cardiac surgery be postponed until 5 to 7 days after cessation of treatment if the patient has been undergoing treatment with ASA.

References

1. Weber, E. et al., Unerwünschte Wirkungen nach Pyrazolonderivaten und Acetylsalicylsäure. In: Kommerell, B. et al. (Eds.), Fortschritte der Inneren Medizin. Springer-Verlag, Berlin 1982, pp. 363–369.
2. Piazolo, A., Zum Auftreten gastrointestinaler Blutungen unter Acetylsalicylsäure als Prophylaktikum thromboembolischer Ereignisse unter Berücksichtigung der Dosisabhängigkeit. Dissertationsschrift, Universität Heidelberg (1989).
3. Bousser, M. G. et al., "AICLA" controlled trial of aspirin and dipyridamole in the secondary prevention of athero-thrombotic cerebral ischemia. Stroke **14**: 5–14 (1983).
4. Carson, J. L. et al., The association of nonsteroidal anti-inflammatory drugs with upper gastrointestinal tract bleeding. Arch. Intern. Med. **147**: 85–88 (1987).
5. Weber, E. et al. (1991), Is gastrointestinal bleeding following the intake of aspirin dose-dependent? In: Brune, K., and Santoso, B. (Eds.), Antipyretic Analgesics: New Insights. Birkhäuser Verlag, Basel 1992, pp. 49–55.
6. UK-TIA Study Group, United Kingdom transient ischaemic attack (UK-TIA) aspirin trial: interim results. Br. Med. J. **296**: 316–320 (1988).
7. Sethi, G. K. et al., Implication of preoperative administration of aspirin in patients undergoing coronary artery bypass grafting. J. Am. Coll. Cardiol. **15**: 15–20 (1990).

Adverse Reactions to NSAID: Clinical Pharmacoepidemiology
ed. by M. Kurowski
© 1992 Birkhäuser Verlag Basel

Gastrointestinal Tract Toxicity – A Risk Factor for Non-Steroidal Anti-Inflammatory Therapy

C. J. Hawkey

Department of Therapeutics, University Hospital, Queen's Medical Centre, Nottingham NG7 2UH, United Kingdom

The gastroduodenal side effects of non-steroidal anti-inflammatory drugs (NSAIDs) have received much attention recently, but there are widely discrepant views regarding the magnitude of risk entailed by taking these drugs [1].

The range of opinion

While it has been claimed that "as a group these drugs are probably the safest used in medicine today", endoscopic studies have revealed the incidence and prevalence of ulcers in patients using NSAIDs. Various studies of patients chronically receiving NSAIDs have put the prevalence of gastric ulcers at between 9% and 22% and the prevalence of duodenal ulcers at between 5% and 22% [1–5]. In three recent studies on the prophylaxis of NSAID ulcers, patients with no ulcers at initial endoscopy who received NSAIDs without prophylactic co-therapy developed gastric ulcers (6–22%) and duodenal ulcers (3.5–8%) over a period of two to three months [6–8]. None of these studies had a control group of patients not taking NSAIDs but their implications are alarming and suggest that patients taking NSAIDs might be 100 times more likely to develop gastric ulcers than those not using these drugs.

Case-control studies

Case-control studies [9–15] present a picture very different from that of the studies cited above (Fig. 1). For example, meta-analysis suggests that the chances of presenting with hematemesis and melena are increased 3.3-fold (95% confidence interval 2.4–4.5) for aspirin and 3.1-fold (95% c.l. 2.3–4.2) for non-aspirin NSAIDs (NANSAIDs) [1]. Moreover, most deaths occur in the elderly. It is with

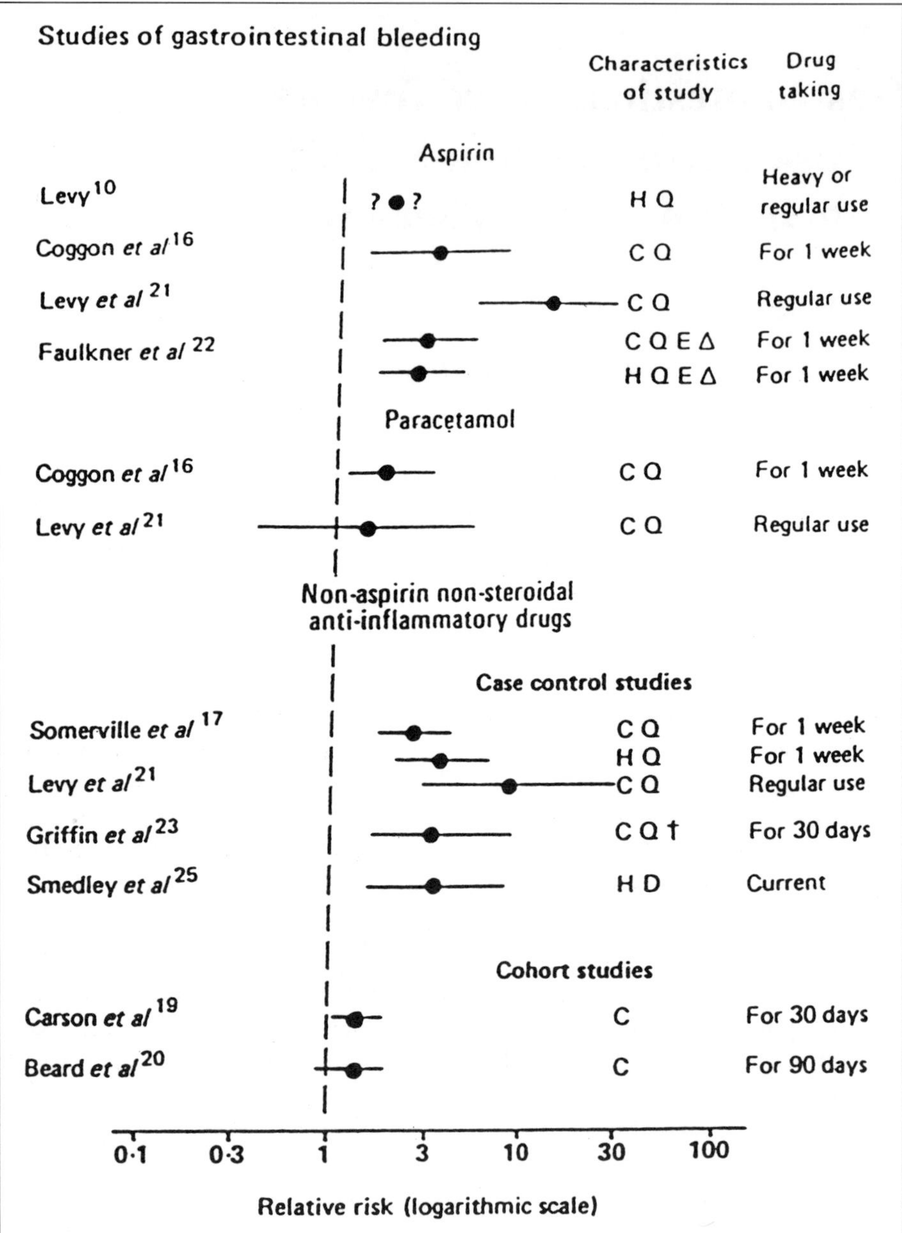

Fig. 1. Relative risks of upper gastrointestinal bleeding for patients taking aspirin, paracetamol, or non-aspirin non-steroidal anti-inflammatory drugs. H = comparison with hospital controls; C = comparison with community controls; Q = quoted values; D = derived values; E = elderly patients (over 60 or 65); △ = ulcer bleeding only; † = death attributable to gastrointestinal bleeding; ? = confidence limits not given.
(Reproduced with kind permission of the Editor, British Medical Journal)

this group that the principal hazard lies, for the elderly consume most NSAIDs. However, even where patients present relatively involuntarily as with hematemesis and melena, it can be argued that case control studies are subject to some bias.

Cohort studies

Cohort studies, in which patients are followed prospectively, are less subject to bias, and two such studies have suggested that NSAIDs only enhance the risk of bleeding by approximately 50% [16,17]. In one of these studies the increase was statistically significant, while in the other study it was not.

A different cohort study [18] (data *not* shown on Fig. 1) which examined both bleeding and perforation of ulcers reported a somewhat higher overall relative risk [2.11-fold (95% c.l. 1.57–2.84) for males and 2.05-fold (95% c.l. 1.63–2.50) for females], which was somewhat lower than the risks computed from case control studies and markedly lower than the risks implied by endoscopic surveys. The relative risks did not show an obvious progression with age although the attributable risk did, i.e., elderly patients are not necessarily more sensitive to NSAIDs but they are more prone to ulceration, a situation exacerbated by their being more likely to consume NSAIDs than younger patients. This study also reported an association between NSAID use and all upper gastrointestinal events (World Health Organization International Classification of Diseases, 9th edition, codes 520–579 and 787). Attributable event rates of 20 to 110 per 1,000 NSAID users over two years (depending on age) reported in this study have been used to suggest that the problem is of greater magnitude than previously recognized [19]. However these WHO ICD codes include trivial lesions and some maladies, such as dental caries, Costen's syndrome, and gallstones, where the association may arise because NSAIDs are a therapy for the condition rather than the cause of it. Moreover, the Office of Population Census and Surveys statistics for ulcer mortality report that in the year 1988 769 people over the age of 60 died of bleeding gastric ulcers and 835 died of bleeding duodenal ulcers in the United Kingdom. If one-third of these deaths are attributable to NSAIDs, then the absolute death rate from bleeding ulcers due to NSAIDs is ca. 500 per annum.

Underestimated risk?

It has been suggested that a higher proportion of NSAID-related deaths occur at home, making the attributable risk derived from hospital-based studies an underestimate [2]. However, it seems unlikely that this inflates the ulcerogenic mortality attributable to NSAID use by more than 25% [21]. This in turn implies that the prognosis of an NSAID-induced ulcer is worse once bleeding has begun, whereas the existing evidence generally runs counter to that view [22].

Low-dose aspirin

There is yet a fourth, albeit ad hoc, cohort study on the risks of NSAIDs, viz. the U. S. Physicians' Study [23]. In this study subjects were randomized to receive 325 mg of aspirin or placebo on alternate days. The incidence of myocardial infarction appeared to be halved in the patients receiving aspirin, but there was a highly significant increase (ca. 50%) in these patients in the rate of presentation with melena. This figure is similar to that reported for full-dose NSAIDs, and raises the possibility either that low-dose aspirin is of comparable toxicity to the gastric mucosa as full-dose aspirin, or that the anti-hemostatic effects of NSAIDs are important when patients present with hematemesis and melena.

What process underlies the risk?

The discrepant estimates of risk discussed above cannot be reconciled if it is assumed that each estimate measures the same processes. It seems highly likely that uncontrolled endoscopic studies detect and classify as ulcers trivial lesions which are not particularly harmful to the patient and which may disappear spontaneously by adaptation to continued NSAID use. It is also of interest that endoscopic studies predominantly detect gastric ulcers, whereas both gastric and duodenal ulcers are equally represented in studies of patients presenting with hematemesis and melena, a proportion similar to that seen in patients not taking NSAIDs. Although fewer studies address this particular matter, non-ulcer bleeding likewise appears to be increased in patients taking NSAIDs [1,11]. These findings, considered jointly with the findings of the U. S. Physicians' Study and with evidence that many episodes of hematemesis and melena occur early during NSAID treatment, raise the possibility that the anti-hemostatic effect of NAN-SAIDs and aspirin is at least partially responsible for presentation with hematemesis and melena. The most extreme view compatible with the available evidence is that lesions detected in acute endoscopic studies are entirely harmless and that presentation with hematemesis and melena occurs solely because the anti-hemostatic actions of NSAIDs cause preexisting chronic silent ulcers to bleed. The implications of this hypothesis for prophylaxis differ from those of the conventional belief that the ulcerogenic effects *per se* of NSAIDs are responsible for presentation with hematemesis and melena.

The truth probably lies somewhere between these two points of view. In order to understand how such risks may arise, it is useful to review the actions of prostanoids and the effects of NSAIDs which may underlie mucosal toxicity and hemostasis.

Mechanism of injury

In 1971 Vane suggested that inhibition of prostaglandin synthesis was the basis of both the therapeutic and adverse effects of NSAIDs. It remains the best unifying hypothesis about the effects of these drugs, but other actions have since been recognized. The protective actions of prostaglandins that are impaired by drugs inhibiting their synthesis include:

1. **Mucus**. The thickness of the mucus coat is enhanced by prostaglandins and impaired by NSAIDs, though whether this is an effect on mucus synthesis or secretion has not been fully established [24].
2. **Surface epithelial bicarbonate secretion**. Duodenal bicarbonate secretion is more substantial in the duodenum than in the stomach [25].
   Between them, mucus and bicarbonate slow acid diffusion four-fold and act as a barrier to macromolecules.
3. **Surface epithelial cell barrier**. The hydrophobicity of the waxy phospholipid cell membrane is impaired by aspirin [26].
4. **Mucosal blood flow**. Vasoconstriction by agents including indomethacin renders injurious concentrations of bile acids that would otherwise be harmless [27].
5. **Cytoprotection**. This term usually describes the phenomenon of mucosal protection by prostaglandins rather than a specific mechanism. Nevertheless, some authors have reported that indomethacin is injurious to isolated human gastric glands and that this injury can be prevented by prostaglandins [28].
6. **Water flux**. Prostaglandins increase serosal mucosal water flux and may "histodilute" injurious agents [29].

Role of prostaglandins

Inhibition of prostaglandin synthesis by NSAIDs is not sufficient to bring about injury to the gastric mucosa. In the absence of topical irritants no injury may arise from the profound inhibition of gastric mucosal prostaglandin synthesis by NSAIDs [30]. Non-topically-induced injury requires inhibition of prostaglandin synthesis but does not correlate with its extent [31]. These relationships are illustrated by salicylates, which are topical irritants. Aspirin inhibits prostaglandin

78

synthesis *and* injures the mucosa; in contrast, sodium salicylate does not inhibit prostaglandin synthesis and is only injurious in the copresence of parenteral indomethacin [32]. In humans, the presence of an acetyl moiety in the salicylate molecule *and* the ability to inhibit prostaglandin synthesis are jointly required for significant injury to occur [33].

Basis of topical irritancy

The basis of topical irritancy is not known. Irritancy is accompanied by electro-physiological changes construed as a manifestation of the breakdown of the mucosal barrier. Underlying events may include disruption of tight junctions, or a metabolic effect on the metabolism of epithelial cells such as uncoupling of oxidative phosphorylation.

Other actions of NSAIDs

NSAIDs have actions whose existence, relationships to prostaglandin synthesis, and importance vis-à-vis the gastric mucosa are less firmly established. These include:
(a) enhancement of oxygen free radical production, possibly as a consequence of reperfusion;
(b) increased production of cytokines such as interleukin-1 [34];
(c) substrate diversion from prostaglandins to more toxic substances such as leukotrienes. This appealing notion has never been clearly demonstrated but is worth reinvestigation in view of:
(d) neutrophil adhesion to blood vessels. This process as well as NSAID-induced injury are prevented in animals by neutropenia, although not by leukotriene antagonists [35].

Effect of NSAIDs on repair processes

Following superficial injury the gastric mucosa reconstitutes itself by a process known as rapid epithelial restitution (RER). RER occurs within minutes or hours and does not involve cell division. *In vitro* studies show that prostaglandins and NSAIDs have no direct effect upon RER; however, NSAIDs and other vasocon-strictors reduce secretion of bicarbonate below the epithelial cap of desquamated fibrin and cells, resulting in a fall in juxtamucosal pH which inhibits RER *in vivo* [36].

Processes other than RER must be involved in ulcer healing. At some stage cell division must occur, and formation of granulation tissue in the base and rims of the ulcer are an important aspect of healing. A number of studies have shown that prostaglandins stimulate cell division [33] and that cell division is acutely inhibited in the absence of prostaglandin synthesis [37]. Unfortunately the picture is confused, since long-term treatment with low doses of indomethacin also results in stimulated cell division and hyperplasia. It seems likely that this is a secondary adaptive response to indomethacin-induced injury.

Other data suggest that prostaglandins play a critical role in new vessel formation in the granulation tissue of ulcers. In indomethacin-treated animals, ulcers which heal have fewer new vessels and are more fibrous and less contractile than control animals not treated with indomethacin [39,40].

Is an anti-hemostatic effect of NSAIDs important?

There are several lines of evidence to support the suggestion that impaired hemostasis contributes to the development of hematemesis and melena in patients taking NSAIDs:

(i) Bleeding often occurs early in the course of treatment with NANSAIDs or after casual aspirin use. Bleeding risk appears to rise to a maximum after four scripts and then declines.

(ii) Bleeding from acute lesions caused by piroxicam becomes significant when serum concentrations of piroxicam reach levels capable of inhibiting platelet function [41].

(iii) Different therapeutic strategies have been shown to reduce the number of acute erosions (enteric coating [42]) or affect their tendency to bleed (changes in pH [43,44]).

(iv) The amount of bleeding induced by mucosal biopsy is greater in subjects taking aspirin compared to control conditions [45].

Synopsis

The above information does not allow one to draw conclusions about the relative roles of mucosal injury and hemostasis in the development of hematemesis and melena. While the greater vascularity of ulcers in animals not given NSAIDs enhances their healing rate, it also makes them more likely to bleed following exposure to an anti-hemostatic agent.

Practical prophylaxis

Four studies have addressed the issue of practical prophylaxis [6–8,46]. Two studies of ranitidine ("Zantac") showed that normal doses have no significant effect on the development of gastric ulceration but are capable of preventing the development of acute duodenal ulcers in patients also taking NSAIDs; in contrast, misoprostol, an anti-ulcerative prostaglandin E_1 analogue, is capable of reducing the incidence of acute gastric ulceration compared both to placebo and to sucralfate while its effects on the duodenum remain uncertain. It should be emphasized that these studies addressed the issue of primary prophylaxis (i.e., prevention of ulcers) developing in patients not previously known to have ulcers. The issue facing the clinician is that of secondary prophylaxis, or how to prevent ulcer recurrence in a patient with an uncomplicated or bleeding ulcer. There are no data clearly addressing this important point. Whatever the truth, the very high frequency of lesions classified as ulcers in acute endoscopic studies raises some doubts about the validity of these studies in identifying proper prophylactic strategies.

References

1. Hawkey, C. J., Non-steroidal anti-inflammatory drugs and peptic ulcers. Facts and figures multiply, but do they add up? Br. Med. J. **300**: 278–284 (1990).
2. Sun, D. C. H., Roth, S. H., Mitchell, C. S., and Englund, D. W., Upper gastrointestinal disease in rheumatoid arthritis. Dig. Dis. **19**: 405–410 (1974).
3. Collins, A. J. and du Toit, J. A., Upper gastrointestinal findings and faecal occult blood in patients with rheumatic diseases taking nonsteroidal antiinflammatory drugs. Br. J. Rheumatol. **26**: 295–298 (1987).
4. Larkai, E. N., Smith, L. J., Lidsky, M. D., and Graham, D. Y., Gastroduodenal mucosa and dyspeptic symptoms in arthritic patients during chronic nonsteroidal antiinflammatory drug use. Am. J. Gastroenterol. **82**: 1153–1158 (1987).
5. Farah, D., Sturrock, R. D., and Russell, R. I., Peptic ulcer in rheumatoid arthritis. Ann. Rheum. Dis. **47**: 478–480 (1988).
6. Ehsanullah, R. S. B., Page, M. C., Tildesley, G., and Wood, J.R., Prevention of gastroduodenal damage induced by non-steroidal anti-inflammatory drugs: controlled trial of ranitidine. Br. Med. J. **297**: 1017–1021 (1988).
7. Graham, D. Y., Agrawal, N., and Roth, S. H., Prevention of NSAID-induced gastric ulcer with the synthetic prostaglandin, misoprostol – a multicenter, double-blind, placebo-controlled trial. Lancet ii: 1277–1281 (1988).
8. Robinson, M. G., Griffin, J. W., Bowers, J. et al., Effect of ranitidine gastroduodenal mucosal damage induced by nonsteroidal antiinflammatory drugs. Dig. Dis. Sci. **3**: 424–428 (1989).
9. Levy, M., Aspirin use in patients with major upper gastrointestinal bleeding and pepticulcer disease. N. Engl. J. Med. **290**: 1158–1162 (1974).
10. Coggon, D., Langman, M. J. S., and Spiegelhalter, D., Aspirin, paracetamol, and haematemesis and melaena. Gut **23**: 340–344 (1982).

11. Somerville, K., Faulkner, G., and Langman, M. J. S., Non-steroidal anti-inflammatory drugs and bleeding peptic ulcer. Lancet i: 462–464 (1986).
12. Levy, M., Miller, D. R., Kaufman, D. W., et al., Major upper gastrointestinal trace bleeding. Relation to the use of aspirin and other nonnarcotic analgesics. Arch. Intern. Med. **148**: 281–285 (1988).
13. Faulkner, G., Prichard, P., Somerville, K., and Langman, M. J. S., Aspirin and bleeding peptic ulcers in the elderly. Br. Med. J. **297**: 1311–1313 (1988).
14. Griffin, M. R., Ray, W.A., and Shaffner, W., Nonsteroidal anti-inflammatory drug use and death from peptic ulcer in elderly persons. Ann. Intern. Med. **109**: 359–363 (1988).
15. Smedley, F. H., Taube, M., Leach, R., and Wastell, C., Non-steroidal anti-inflammatory drugs: retrospective study of bleeding and perforated peptic ulcers. Gut **29**: A1443 (1988).
16. Carson, J. L., Strom, B. L., Soper. K. A., West. S. L., and Morse, M. L., The association of non-steroidal anti-inflammatory drugs with upper gastrointestinal tract bleeding. Arch. Intern. Med. **147**: 85–88.
17. Beard, K., Walker. A. M., Perera. D. R., and Jick, H., Non-steroidal anti-inflammatory drugs and hospitalization for gastroesophageal bleeding in the elderly. Arch. Intern. Med. **147**: 1621–16.23 (1987).
18 Beardon, P. H. G., Brown. S. V., and McDevitt, G., Gastrointestinal events in patients prescribed non-steroidal anti-inflammatory drugs: A controlled study using record linkage in Tayside. Q. J. Med. **266**: 497–505 (1989).
19. Shield, M. J., NSAIDs and peptic ulcers. Br. Med. J. **300**: 814–816 (1990).
20. Armstrong, C. P. and Blower, A. L., Non-steroidal anti-inflammatory drugs and life threatening complications of peptic ulceration. Gut **28**: 527–532 (1987).
21. Quader, K. and Logan, R. F. A., Peptic ulcer (PU) deaths: how many occur at home or after non-steroidal anti-inflammatory drug (NSAID) prescribing? Gut **29**: A1443 (1988).
22. Henry, D. A., Johnson. A., Dobson, A., and Duggan, J., Fatal peptic ulcer complications and the use of non-steroidal, anti-inflammatory drugs, aspirin and corticosteroids. Br. Med. J. **295**: 1227–1229 (1987).
23. The Steering Committee of the Physicians' Health Study Research Group, Final report on the aspirin component of the ongoing Physicians' Health Study. N. Engl. J. Med. **321**: 129–135 (1989).
24. McQueen, S., Hutton. D., Allen. A., and Garner, A., Gastric and duodenal surface mucus gel thickness in rat: effects of prostaglandins and damaging agents. Am. J. Physiol. **245**: G388–G393 (1983).
25. Hogan, D. L., and Isenberg, J. I., Gastroduodenal bicarbonate production. Adv. Intern. Med. **33**: 385–408 (1988).
26. Lichtenberger, L. M., Richards. J. E., and Hills, B. A., Effect of 16,16-dimethylprostaglandin E_2 on surface hydrophobicity of aspirin-treated canine gastric mucosa. Gastroenterology **88**: 308–314 (1985).
27. Whittle, B. J. R., Mechanisms underlying gastric mucosal damage by indomethacin and bile salts and the action of prostaglandins. Br. J. Pharmacol. **60**: 455–460 (1977).
28. Tarnawski, A., Brzozowski, T., Sarfeh, I. J. et al., Prostaglandin protection of human isolated gastric glands against indomethacin and ethanol injury. Evidence for direct cellular action of prostaglandin. J. Clin. Invest. **81**: 1081–1089 (1988).
29. Pihan, G. and Szabo, S., 16,16-$DMPGE_2$ and thiosulfate decrease gastric mucosal penetration of hypertonic NaCl and increase net transmucosal water flux. Gastroenterology **94**: A354 (1988).
30. Ligumsky, M., Golanska, E. M., Hansen, D. G., and Kauffman, G. L., Jr., Aspirin can inhibit gastric mucosal cyclooxygenase without causing lesion in the rat. Gastroenterology **84**: 756–761 1983).

31. Ligumsky, M., Sestieri, M., Karmeli, F., Zimmerman, J., Okon, E., and Rachmilewitz, D., Rectal administration of nonsteroidal antiinflammatory drugs. Effect on rat gastric ulcerogenicity and prostaglandin E_2 synthesis. Gastroenterology **98**: 1245–1249 (1990).

32. Steele, G. and Whittle, B. J. R., Gastric damage induced by topical salicylate can be potentiated by aspirin or indomethacin. Br. J. Pharmacol. **81**: 78P (1984).

33. Mahida, Y. R., Bhaskar, N. K., Hurst, S., Daneshmend, T. K., and Hawkey, C. J., Comparative gastric mucosal toxicity of different salicylates in humans. Gut **31**: A598 (1990).

34. Knudsen, P. J., Dinarello, C. A., and Strom, T. B., Prostaglandins posttranscriptionally inhibit monocyte expression of interleukin-1 activity by increasing intracellular cyclic adenosine monophosphate. J. Immunol. **137**: 3189–3194 (1986).

35. Wallace, J. L., Keenan, C. M., and Granger, D. N., NSAID-induced gastropathy: a neutrophil dependent process. Gastroenterology **98**: A145 (1990).

36. Wallace, J. L and McKnight, G. W., The mucoid cap over superficial gastric damage in the rat. A high-pH microenvironment dissipated by nonsteroidal antiinflammatory drugs and endothelin. Gastroenterology **99**: 295–304 (1990).

37. Levi, S., Goodlad, R. A., Lee, C. Y., Stamp, G. et al., Inhibitory effect of nonsteroidal anti-inflammatory drugs on mucosal cell proliferation associated with gastric ulcer healing. Lancet **336**: 840–843 (1990).

38. Goodlad, R. A., Madgwick, A. J., Moffatt, M. R., Levin, S., Allen, J. L., and Wright, N. A., Effects of misoprostol on cell migration and transit in the dog stomach. Gastroenterology **98**: 90–95 (1990).

39. Tarnawski, A., Stachura, J., Douglass, T. G., Krause, W. J., Gergely, H., and Sarfen, I. J., Does indomethacin affect quality of experimental gastric ulcer healing? A quantitative histologic and ultrastructural analysis. Gastroenterology **98**: A136 (1990).

40. Ogihara, Y., Fuse, Y., and Okabe, S., Effects of indomethacin and prednisolone on connective tissue at the base of acetic acid-induced gastric ulcers in rats. In: Garner, A. and O'Brien, P. E., Eds., Mechanisms of injury, protection and repair of the upper gastrointestinal tract. John Wiley & Sons, New York 1991, pp. 455–465.

41. Fellows, I. W., Bhaskar, N. K., and Hawkey, C. J., The nature and time course of piroxicam-induced injury to human gastric mucosa. Aliment. Pharmacol. Therap. **3**: 481–488 (1989).

42. Hawthorne A. B., Hurst S. M., Mahida Y. R., Cole, A. T., and Hawkey, C. J., Aspirin-induced gastric mucosal damage: prevention by enteric-coating of aspirin and relation to prostaglandin synthesis. Br. J. Clin. Pharacol., **30**: 187–194 (1990).

43. Cole, A. T., Brundell, S., Hudson, N., Hawthorne, A. B., and Hawkey, C.J., High dose ranitidine prophylaxis of gastric haemorrhagic lesions. Gut **31**: A1205 (1990).

44. Mann, S. G., Didcote, S., Hyman-Taylor, P., and Hawkey, C.J., Prolongation of intragastric bleeding by acid. Gut **31**: A1206 (1990).

45. Hawkey, C. J., Sharma, H. K., Bhaskar, N. K., Didcote, S. M., Hawthorne, A. B., and Daneshmend, T. K., High and low dose aspirin: equal gastric damage but impaired haemostasis at high dose. (Gut 1989; 30: A1442)

46. Agrawal, N., Stromatt, S., and Brown, J., Comparative study of misprostol and sucralfate in the prevention of NSAID-induced gastric ulcers. Gastroenterology **98**: A14 (1990).

Incidence of Mild and Severe Skin Reactions Associated with the Use of Non-Steroidal Anti-Inflammatory Drugs

E. Schöpf and B. Rzany

*Department of Dermatology, University of Freiburg, Hauptstraße 7,
D-W-7800 Freiburg, Germany*

Non-steroidal anti-inflammatory drugs (NSAIDs) are among the most commonly-prescribed drugs in Germany, and the tendency to prescribe NSAIDs is increasing. In 1988 354,000,000 defined daily doses (DDD) of NSAIDs were prescribed in West Germany, a 6.1% increase over the 1987 figure. A great variety of NSAIDs are known (propionic acid derivatives, oxicam derivatives, arylacetic acid derivatives, nicotinic acid derivatives, acrylacetic acid derivatives, anthranilic acid derivatives, indoleacetic acid derivatives, and pyrazolone derivatives), but the market is controlled by only a few products. The most frequently-prescribed NSAID was the arylacetic acid derivative diclofenac (50% of all prescriptions), followed in descending order by indomethacin, ibuprofen, and piroxicam [1]. NSAIDs of all groups are well known for their various cutaneous side-effects [2]. The ability to cause photosensitivity (i.e. phototoxic reactions) is recognized for the phenylpropionic acid derivatives benoxaprofen, carprofen, ketoprofen, tiaprofenic acid, and naproxen as well as the oxicam derivate piroxicam [3].

Most of the literature on cutaneous side effects of NSAIDs consists of case reports; a notable exception is Stern's paper [2] on the speciality-based system for spontaneous reporting of adverse reactions (ARs) to NSAIDs. In this study a variety of cutaneous reactions were documented: vesiculobullous reactions, toxic epidermal necrolysis (TEN), erythema multiforme, erythroderma exfoliativa, anaphylaxis, angioedema, urticaria, serum sickness, vasculitis, purpura petechiae, photosensitivity, fixed drug eruptions, exanthema, pruritus with and without rash, skin pain, nail disorders, lichenoid reactions, granulomatous eczematous reactions, and alopecia. Cutaneous ARs occurred most commonly with piroxicam, zomepirac sodium, sulindac, meclofenamate sodium, and benoxaprofen [2].

The list of cutaneous ARs associated with NSAIDs may be extended by reports of syringomatous hyperplasia and eccrine squamous syringometaplasia [4], hypersenisitivity angiitis [5], pseudoporphyria [6,7], generalized eruptive pustular drug rash [8], a dermatomyositis-like syndrome [9], and a rash with pulmonary eosinophilia [10].

Although most reactions are self-limited and without sequelae [2], NSAIDs are also reported to be associated with the severe skin reactions TEN [2,11–24] and Stevens-Johnson syndrome (SJS) [24]. Both TEN and SJS are characterized by acute onset, fever, mucosal involvement, and target lesions (SJS) or extensive blistering (TEN) [21,25]. TEN is characterized by a ca. 30% mortality rate [23,26].

As with all adverse drug reactions, it is critical to establish whether certain NSAIDs are associated with an increased risk for inducing TEN and SJS. Following the 1985 discovery in France of several cases of TEN associated with the oxicam derivative isoxicam [20,22], epidemiological studies were performed in France [23] and Germany [26] to determine the incidence of TEN and SJS, the demographic characteristics of these conditions, and the identity of the drugs and classes of drugs involved.

Methods

Medical centers in West Germany considered likely to treat severe skin reactions were contacted. These included dermatology departments, units equipped with facilities for treating severe burns, and intensive care units devoted to general or specialized medical care (but excluding pediatric wards). A total of 1,139 medical centers were contacted by letter; many letter-contacts were followed up by telephone calls and personal visits. All severe skin reactions occurring between 1981 and 1985 that required hospitalization were documented, according to information provided by participating institutions in response to questionnaires.

TEN and SJS were defined in accordance with criteria established by an international group of dermatologists [25]. Drugs causally related to TEN and SJS were defined as those which had "been taken close enough prior to the onset of any symptom (i.e. 21 days)" and believed to reveal a "definite, probable, possible" relationship to the condition. If the AR regressed during continued administration of the drug, a causal relationship between the drug and the AR was deemed unlikely; such drugs were regarded as playing only a "conditional or doubtful" role in the etiology [27]. If, in a single patient, several drugs came under suspicion, they were all taken into account. Assessment of incidences of SJS or TEN caused by medication was based on defined daily doses (DDDs) of drugs sold by pharmacies during the years 1981–1985. A DDD is defined as "the mean dosage of a drug for the main indication given to a 70 kg adult patient" [1].

The overall chance of developing either TEN or SJS (incidence rate for TEN or SJS, IR) was calculated according to the formula

$$IR = D/(P)(C)(Y) ,$$

where

D = number of cases of TEN or SJS diagnosed
P = mean population of West Germany from 1981 through 1985 inclusive
C = fraction coverage (fraction of TEN or SJS cases documented)
Y = number of years investigated .

Results

With coverage rates estimated to be 91% for TEN and SJS (i.e., C = 0.91), 259 cases of TEN and 315 cases of SJS were identified. Using values of 61,400,000 for P and 5 for Y, incidence rates were calculated as 0.93 and 1.1 per 1,000,000 persons per year for TEN and SJS, respectively.

A history of treatment with a drug having a "definite, probable, possible" relationship to the skin reaction could be established in 89% of patients with TEN and 54% of patients with SJS. No association with previous medication could be traced for 31% of the SJS cases and 3% of the TEN cases.

Antibiotics were the most common causative drugs for TEN (42%) and SJS (34%), followed by analgesics (TEN, 23%; SJS, 33%). NSAIDs were causative for 19% of TEN cases and 14% of SJS cases (Tab. 1).

Table 1. Comparison of the main medication groups for drug-induced TEN and SJS.[*]

Disease	TEN	SJS
Patients with drug-induced TEN/SJS	216	164
Causative drug class	**% of cases**	**% of cases**
Antibiotics	42	34
Analgesics	23	33
NSAIDs	19	14
Anti-Gout drugs	15	5
Psycholeptics	11	7
Cough and cold preparations	6	21

* A case may be included in more than one medication group.

Table 2. Sales (per 1 million DDDs), number of cases of TEN (N), and incidences of TEN (per 1 million DDDs).

Drug ingredient	Sales	N	Incidence
Benoxaprofen	4	1	0.25
Phenylbutazone, dexamethasone, propyphenazone, thiamine, orphenadrine citrate	5	1	0.22
Lonazolac	8	1	0.14
Phenylbutazone	21	2	0.11
Oxyphenbutazone	30	2	0.07
Phenylbutazone, metamizol, diphenhydramine, aneurine, cyanocobalamine	32	2	0.07
Isoxicam	28	1	0.04
Tiaprofenic acid	30	1	0.04
Diclofenac	507	12	0.03
Piroxicam	219	5	0.03
Indomethacin	414	7	0.02
Ketoprofen	84	1	0.01

Considering only drugs or drug combinations with DDDs of more than 2,500,000 the highest incidences for TEN were found for some sulfonamides and certain other antibiotics. For NSAIDs, incidences ranged from 0.01–0.25/1,000,000 DDD in TEN patients (Tab. 2) and from 0.005–0.25/1,000,000 DDD in SJS patients (Tab. 3).

Discussion

Case reports as well as studies by Stern [2] and Roujeau [23] indicate that NSAIDs definitely contribute to the incidence of TEN and SJS. Previous case reports on TEN associated with benoxaprofen [2,17,18], piroxicam [22,23], isoxicam [20,22,23], indomethacin [23,28], phenylbutazone [12–16,22], and oxyphenbutazone [15,19,21–24] were confirmed by our study. The NSAIDs ketoprofen, tiaprofenic acid, lonazolac, and propyphenazone have now also been found to be associated with TEN. Drugs not included in our study which have been implicated as causative agents of TEN include flurbiprofen [22], diclofenac [23,29], zomepirac sodium [2,30], niflumic acid [22], tolmetin [2], fenbufen [22,23], and sulindac [2].

Table 3. Sales (per 1 million DDDs), number of cases of SJS (N), and incidences of SJS (per 1 million DDDs).

Drug ingredient	Sales	N	Incidence
Dexamethasone, salicylamide, mofebutazone, magnesium gluconate, trimethylhesperidin chalcone	13	3	0.25
Benoxaprofen	4	1	0.24
Phenylbutazone, dexamethasone, propyphenazone, thiamine, orphenadrine citrate	5	1	0.21
Isoxicam	28	4	0.15
Oxyphenbutazone, propyphenbutazone	15	2	0.15
Phenylbutazone	21	2	0.10
Phenylbutazone, metamizol, diphenhydramine, aneurine, cyanocabalamine	32	3	0.10
Oxyphenbutazone	30	2	0.07
Phenylbutazone, dexamethasone, cyanocobalamin, aneurine chloride, aluminium glycinat	38	1	0.03
Piroxicam	219	3	0.01
Indomethacin	414	2	0.01
Diclofenac	507	2	< 0.005

Incidences calculated from DDD data should be cautiously interpreted, with comparisons restricted to drugs within a single drug group rather than between drugs in different NSAID classes. For drugs belonging to the same NSAID class we found differences in the incidences for TEN and SJS of up to 25/1,000,000 DDDs, whereas Roujeau [23] reported differences as great as 38/1,000,000 DDDs. Within a class of closely-related drugs a difference in incidences of more than 10/1,000,000 DDDs was thought to be significant. In France, isoxicam was withdrawn from the market following a series of case reports on its association with TEN [20] and on the basis of a difference of 10 for incidences of TEN for isoxicam compared to the other oxicam derivative (piroxicam) on the market. In the German study there was no difference in the incidence of TEN per 1,000,000 DDDs between isoxicam and piroxicam (Table 4).

Conclusion

There is no doubt that NSAIDs contribute to a variety of skin reactions, including the severe reactions known as TEN and SJS. As calculations of incidences per

88

Table 4. TEN associated with drugs: comparison of sale numbers per million DDDs (sale no.), numbers of cases where the drugs were mentioned (no. cases) and incidences of TEN per million DDDs (inc. no.) between the German [26] and French [23] studies for the years 1981–1985.

Drug groups	Schöpf et al.			Roujeau et al.		
	Sale no. in DDD	Abs. no. of cases	Inc. per 1 mill. DDD	Sale no. in DDD	Abs. no. of cases	Inc. per 1 mill. DDD
Antibiotics						
Sulfonamides						
– Pyrimethamine, Sulfadoxine	6	6	1.14	–	–	–
– Cotrimoxazol	153	34	1.25	203	21	0.103
β-Lactams						
– Ampicillin	12	8	0.73	85	8	0.094*
– Amoxicillin	25	5	0.22	342	12	0.035
Tetracyclines						
– Doxycycline	205	3	0.09	–	–	–
NSAIDS						
– Benoxaprofen	4	1	0.25	–	–	–
– Oxyphenbutazone	30	2	0.07	112	18	0.1607
– Isoxicam	28	1	0.04	34	13	0.382
– Diclofenac	507	12	0.03	524	6	0.014
– Piroxicam	219	5	0.03	371	13	0.035
– Indomethacin	414	7	0.02	272	3	0.01
Anti-Gout drugs						
– Allopurinol	926	30	0.04	837	7	0.013*
Psycholeptics						
– Phenytoin	199	10	0.06	43	4	0.14*
– Carbamazepine	103	4	0.04	99	6	0.06
Analgesics						
– Paracetamol	165	3§	0.02	1800	27	0.015*
– ASA	398	3§	0.01	3600	13.1	0.008*

*incidences were calculated for every case when a drug was mentioned.
§ sum of all drugs and drug combinations where substance was included.

1,000,000 DDDs show differences up to 25 [26] or 38 [23], some NSAIDs may be associated with an elevated risk. Due to other ARs the NSAIDs with the highest incidences of TEN and SJS per 1 million DDDs in the German study (benoxaprofen, oxyphenbutazone) have *already* been withdrawn from the market. Severe adverse skin reactions may be a first sign of toxicity of a drug and thus should be cautiously monitored.

Summary

A retrospective epidemiological study was conducted in West Germany with the aim of collecting all hospitalized cases of toxic epidermal necrolysis (TEN) and Stevens-Johnson syndrome (SJS) between 1981 and 1985. Two hundred and fifty-nine cases of TEN and 315 cases of SJS were identified; with coverage rates estimated at 91%, these case loads correspond to incidence rates of 0.93 per 1,000,000 persons per year for TEN and 1.1 per 1,000,000 persons per year for SJS. A history of treatment with a drug having a "definite, probable, [or] possible" relationship to the skin reaction could be established in 89% of the patients with TEN and 54% of those with SJS. Antibiotics were involved in 42% of TEN cases and 34% of SJS cases, whereas analgesics were involved in 23% of TEN cases and 33% of SJS cases. NSAIDs accounted for 19% of the TEN cases and 14% of the SJS cases. In patients treated with NSAIDs, incidences of TEN and SJS ranged from 0.1–0.25 per 1,000,000 defined daily doses (DDD) and 0.005–0.25 per 1,000,000 DDD, respectively.

References

1. Schwabe, Paffrath, Arzneiverordnungsreport '89. Gustav Fischer Verlag, Stuttgart 1989, pp. 108–116.
2. Stern, R. S. and Bigby, M., An expanded profile of cutaneous reactions to nonsteroidal anti-inflammatory drugs. J. Am. Med. Assn. **252**: 1433–1437 (1984).
3. Kochevar, I. E., Phototoxicity of nonsteroidal inflammatory drugs. Arch. Dermatol. **125**: 824–826 (1989).
4. Lerner, T. H., Barr, R. J., Dolezal, J. F., and Stagone, J. J., Syringomatous hyperplasia and eccrine squamous syringometaplasia associated with benoxaprofen therapy. Arch. Dermatol. **123**: 1202–1204 (1987).
5. Singhal, P. C., Faulkner, M., Venkatesan, J., and Molho L., Hypersensitivity angiitis associated with naproxen. Ann. Allergy, **63**: 107–109 (1989).
6. Rivers, J. K. and Barnetson, R. S., Naproxen-induced bullous photodermatitis. Med. J. Aust. **151**: 167–168 (1989).
7. Suarez, S. M., Cohen, P. R., and DeLeo, V. A., Bullous photosensitivity to naproxen: "pseudopophyria". Arthritis Rheum. **33**: 903–908 (1990).
8. Grattan, C.E., Generalised eruptive pustular drug rush due to naproxen. Dermatologica, **179**: 57–58 (1989).
9. Grob, J. J., Collet, A. M., and Bonerandi, J. J., Dermatomyositis-like syndrome induced by nonsteroidal anti-inflammatory agents. Dermatol., 178, 58–59 (1989).
10. Burton, G. H., Rash and pulmonary eosinophilia associated with fenbufen. Br. Med. J. **12**: 82–83 (1990).
11. Lyell, A., Toxic epidermal necrolysis: An eruption resembling scalding of the skin. Br. J. Dermatol. **68**: 355–361 (1956).
12. Oventon, J., Toxic epidermal necrolysis associated with phenylbutazone therapy. Br. J. Dermatol. **74**: 100 (1962).

13. Vas, C. J., Unusual complications of phenylbutazone therapy – toxic epidermal necrolysis. Postgrad. Med. J. **39**: 94 (1963).
14. Ohlenschleger, K., Toxic epidermal necrolysis and Stevens-Johnson's disease. Acta Derm. Venereol. **46**: 204–209 (1966).
15. Lyell, A., A review of toxic epidermal necrolysis in Britain. Br. J. .Dermatol. **79**: 662–671 (1967).
16. Montgomery P. R., Toxic epidermal necrolysis due to phenylbutazone. Br. J. Dermatol. **83**: 220.
17. Fenton, D. and English, J. S., Toxic epidermal necrolysis, leucopenia and thrombocytopenic purpura–a further complication of benoxaprofen therapy. Clin. Exp. Dermatol. **7**: 277–279 (1982).
18. Palframan, A. and Makepeace, A. R., Side effects of benoxaprofen. Br. Med. J. **285**: 376–377 (1982).
19. Broekhuizen, T. H., Nieuborg, L., and Dinkelman, R.J., Homograft as biological dressing in toxic epidermal necrolysis. Lancet 2, 1023–1024 (1983).
20. Flechet, M. L., Moore, N., Chedeville, J. C., Paux, G., Boismare, F., and Lauret, P. (1985), Fatal epidermal necrolysis associated with isoxicam. Lancet 2, 499 (1985).
21. Ruiz-Maldonado, R., Acute disseminated epidermal necrolysis 1, 2, 3: Study of sixty cases. J. Am. Acad. Dermatol. **13**: 623–635 (1985).
22. Guillaume, J. C., Roujeau, J. C., Revuz, J., Penso, D., and Touraine, R., The culprit drugs in 87 cases of toxic epidermal necrolysis (Lyell's syndrome). Arch. Dermatol. **123**: 1166–1170 (1987).
23. Roujeau, J. C., Guillaume, J. C., Fabre, J. P., Penso, D., Flechet, M. L., and Girre, J. P., Toxic epidermal necrolysis (Lyell syndrome). Incidence and drug etiology in France, 1981–1985. Arch. Dermatol. **126**: 37–42 (1990).
24. Alanko, K., Stubb, S., and Kaupinnen, K., Cutaneous drug reactions: clinical types and causative agents. Acta Derm. Venereol. **69**: 223–226 (1989).
25. Chan, H. L., Stern, R. S., Arndt, K. A., Langlois, J., Jick, S. S., Jick, H., and Walker, A. W. (1990), The incidence of erythema multiforme, Stevens-Johnson syndrome, and toxic epidermal necrolysis. Arch. Dermatol. **126**: 43–47 (1990).
26. Schöpf, E., Stühmer, A., Rzany, B., Victor, N., Zentgraf, R., and Kapp, J. F., Toxic epidermal necrolysis (TEN) and Stevens-Johnson syndrome (SJS). An epidemiological study conducted in West Germany. Arch. Dermatol. **127**: 839–842 (1991).
27. Karch, F. E. and Lasagna, L., Adverse drug reactions. A critical review. J. Am. Med. Assn. **234**: 1236–1241 (1975).
28. O'Sullivan, M., Hanly, J. G., and Molloy, M., A case of toxic epidermal necrolysis secondary to indomethacin. Br. J. Rheumatol. **22**: 47–49 (1983).
29. Kamanabroo, D., Schmitz-Landgraf, W., and Czarnetzki, B. M., (1985), Plasmapheresis in severe drug-induced toxic epidermal necrolysis. Arch. Dermatol. **121**: 1548–1549 (1985).
30. Levitt, L. and Pearson, R. W., Sulindac-induced Stevens-Johnson toxic epidermal necrolysis syndrome. J. Am. Med. Assn. **243**: 1262–1263 (1980).

Kidney Function Impairment: Do Relevant Differences between Non-Steroidal Anti-Inflammatory Drugs Exist?

J. Mann and M. Goerig

4. Medizinische Klinik, Universität Erlangen-Nürnberg, Kontumazgarten 14–18, D-W–8000 Nürnberg 80, Germany

Introduction

Non-steroidal anti-inflammatory drugs (NSAIDs) include such diverse substances as salicylates (aspirin, sodium salicylate, diflunisal, trisalicylate), pyrazoles (phenylbutazone, oxyphenbutazone, azapropazone, feprazone), indene derivatives (indomethacin, sulindac, tolmetin, zomepirac), propionic acid derivatives (fenoprofen, flurbiprofen, ibuprofen, naproxen), fenamates (mefenamic acid, flufenamic acid), oxicams (piroxicam), and acetic acid derivatives (diclofenac, fenclofenac). The therapeutic effectiveness and a substantial part of the adverse effects (AEs) of NSAIDs have been attributed principally to their ability to decrease prostanoid formation by inhibition of the cyclooxygenase activity of prostaglandin endoperoxide synthase (EC 1.14.99.1), the initial enzyme in the synthetic pathway for prostaglandins (PGs) and thromboxanes (TXs) [1–3]. A good correlation exists between the rank order of potencies for inhibiting prostaglandin formation and the reduction of experimental edema *in vivo* [4]. Prostanoids sensitize peripheral nociceptors in the presence of a definite degree of tissue damage, which may explain why NSAIDs are not analgesic in the absence of inflammation.

There are profound and mostly still-unexplained differences in the sensitivity of diverse biological systems to various NSAIDs and in the pattern as well as the probability of side-effects. These differences might be caused by different drug-specific mechanisms of inhibition of cyclooxygenase. Subtle interindividual or tissue-specific variations of cyclooxygenase structure and susceptibility towards NSAIDs, or different rates of penetration by different NSAIDs to the correct active site of the enzyme, or different rates of metabolism of different NSAIDs might also account for the variety of bioresponses to NSAIDs. In this context, it is noteworthy that aspirin binds irreversibly to the active site of cyclooxygenase, whereas indomethacin, ibuprofen, and ibuprofen analogues are competitive inhibitors of cyclooxygenase and can protect the enzyme from deactivation by salicylates [5].

In vitro results suggest that different NSAIDs have different biological effects besides cyclooxygenase inhibition. NSAIDs exhibit substantial differences in their ability to scavenge oxidants or to prevent their formation [6], they exert differential effects on neutrophil function (superoxide anion generation, lysozyme and arachidonic acid release) and plasma membrane viscosity [7], and appear to interfere specifically with post-receptor signalling events such as G-protein function [8,9], protein kinase activity [10], and calcium release from inositol triphosphate-independent intracellular storage pools [11]. However, the clinical relevance of these observations remains to be established.

This short review on adverse renal reactions to NSAIDs focuses on clinically-relevant differences between various NSAIDs. After introducing the main renal effects attributed to PGs, renal syndromes associated with NSAIDs will be considered, followed by an analysis of the different renal effects of various NSAIDs.

Renal effects of PGs

Prostanoids directly influence renal vasculature, renal excretory functions, and renin release [12,13] (Tab. 1). They exhibit complex interactions with the renin-angiotensin-aldosterone system and the kinins, and appear to regulate the synthesis and secretion of atrial natriuretic peptide [14]. Since prostanoids synergize with other inflammatory mediators, their removal reduces the effectiveness of substances such as histamine or bradykinin [4]. The pattern of locally-synthesised prostanoids exhibits a specific intrarenal distribution [12], and the amount of prostanoid synthesis appears to depend strictly on the relative availability of glutathione as a cofactor.

Table 1 lists the main sites of renal PG synthesis and actions. The available evidence indicates that PGs produced in the cortex regulate cortical function and PGs of the medulla act on medullary function, but cortical PGs do not act on the medulla and vice versa. This separation of action is ascribed to regional heterogeneity of PG synthesis and the separate vascular supply of the renal cortex and renal medulla.

PGE_2 and PGI_2 are potent renal vasodilators. In healthy sodium-replete euvolemic subjects vasodilator PGs have little influence on renal function; however, when the activity of renal vasoconstrictors such as catecholamines

Table 1. Major effects and sites of action of prostaglandins in the kidney*

Site	PG	Action
Vasculature	I_2	Vasodilation
Glomerulus	I_2, E_2	Maintains GFR
	TXA_2	Reduces GFR
Collecting tubule	E_2, $F_{2\alpha}$	Enhances NaCl and H_2O excretion
Medullary interstitial cells	E_2	Vasodilation and natriuresis

* adapted from [12]

and angiotensin II is prevailing (as in shock or dehydration), PGE_2 and PGI_2 physiologically antagonize vasoconstrictor activity and contribute to the maintainance of renal perfusion and filtration. In addition, PGE_2 has diuretic and natriuretic properties in the medulla. In the cortex TXA_2 is a vasoconstrictor which reduces the glomerular filtration rate (GFR). This eicosanoid has been implicated in the pathogenesis of glomerulonephritis and acute renal transplant rejection [15a].

Renal syndromes associated with NSAIDs

The main unwanted structural and functional effects of NSAIDs are listed in Table 2. *Interstitial nephritis* has been reported at least for fenoprofen, indomethacin, naproxen, tolmetin, piroxicam, and sulindac [2,16,17]. This renal inflammatory disease differs from the ordinary drug-induced allergic interstitial nephritis in that (a) it is often associated with a nephrotic syndrome, and (b) systemic signs of allergy (eosinophilia, skin rash, fever) are uncommon. However,

Table 2. Renal adverse effects of NSAIDs

Functional adverse effects
Acute and chronic renal insufficiency
NaCl and water retention
Hyperkalemia
Inhibition of diuretic and antihypertensive drug action
Structural adverse effects
Interstitial nephritis
Interstitial nephritis with nephrotic syndrome
Analgesic nephropathy

case reports of systemic illnesses such as tubulointerstitial nephritis with uveitis (TINU syndrome) [17], have appeared. The nephrotic syndrome is due to changes in the glomerular basement membrane indistinguishable from minimal change disease (lipoid nephrosis). In rare cases minimal change lesions in the absence of interstitial inflammation have been described. Interstitial nephritis has been observed 0.5–18 months after initiation of NSAID therapy. It usually resolves within several weeks to several months after withdrawal of the offending drug. It has been proposed that cyclooxygenase inhibition may shunt arachidonic acid to the lipoxygenase pathway with subsequent T cell activation by eicosatetraenoate metabolites. T cells would then be responsible for minimal change disease of the glomeruli and for interstitial cell infiltration [2].

A second structural renal disorder associated with many NSAIDs is *analgesic nephropathy*, which accounts for 5–20% of our hemodialysis population. Risk factors for this disease include being of the female sex, use of combination analgesics, and polytoxicomania. The prevalance of analgesic nephropathy is higher in urban areas than in rural areas, and shows remarkable regional variation. A recent survey indicates that phenacetin and acetaminophen, but not aspirin, are clearly associated with chronic renal failure [18,19]. There is evidence that other non-salicylate NSAIDs may also rarely lead to analgesic nephropathy [20]. Caffeine, a frequent additive to analgesic mixtures, had been previously thought to be only a stimulant for continued drug intake, though this adenosine antagonist is now itself considered a possible toxic substance for the kidney [19]. Adenosine reduces oxygen consumption in several parts of the tubular system; inhibition of adenosine could lead to an imbalance between oxygen supply and tubular metabolic activity, leading to cell death.

Functional acute renal failure due to NSAIDs (renal failure without initial structural damage to the kidney) is frequently encountered in nephrology clinics. As a rule, this side effect does not affect healthy subjects (implying that long-distance running while taking NSAID medication is unhealthy). A number of clinical risk factors for NSAID-induced renal functional impairment have been identified (Tab. 3); they are characterized by low circulatory volume and/or low cardiac output. Vasopressor systems are usually very active in these situations; the kidney is thus protected by PGs from overvasoconstriction. Inhibition of vasodilatory PGs by NSAIDs exposes the kidney to unopposed vasoconstriction, a classic cause of acute renal failure.

Contrary to common believe, acute renal failure can be an immediate rather than a late consequence of NSAID administration; patients at risk for NSAID-induced renal impairment (Tab. 3) must therefore be closely observed shortly after initiation of NSAID therapy. If a rise in serum creatinine occurs, withdrawal of NSAID and administration of fluid usually reduces creatinine levels to baseline values within days.

Table 3. Functional renal failure during NSAID treatment: predisposing factors.

Reduced effective plasma volume Diuretic treatment Low-sodium diet Diarrhea, vomiting Shock (including septic shock) Liver cirrhosis with ascites Pancreatitis
Low cardiac output Congestive heart failure Anesthesia
Renal disease including nephrotic syndrome
Reduced renal function including aged "normal" kidney

Hyperkalemia is another unwanted effect of NSAIDs. It has been attributed to the suppression of renin release and subsequent diminution of aldosterone secretion (hyporeninemic hypoaldosteronism) [20a]. NSAID-induced hyperkalemia occurs especially in patients predisposed for this electrolyte disorder (elderly, low fluid intake, azotemia, intake of potassium-sparing diuretics, intake of angiotensin converting enzyme (ACE) inhibitors, diabetes).

Sodium retention is likewise a common consequence of NSAID treatment [2]. The anti-natriuretic effect of NSAIDs parallels their inhibitory action on renal PG production [21], suggesting a cause-effect relationship. The anti-natriuresis of aspirin and indomethacin in humans is independent of changes in whole kidney blood flow and glomerular filtration [21a], indicating a tubular site of action. NSAIDs also blunt the action of diuretics [22]. Sodium retention in individual patients on continued NSAID treatment may be massive, but mild edema formation is by no means a rare event [2].

Water metabolism may be affected by NSAIDs, since PGs interfere with the diluting and concentrating capability of the kidney. NSAIDs can reduce free water clearance, in part by enhancing the action of anti-diuretic hormone [2]. The latter action is exploited in patients with central and nephrogenic diabetes insipidus. When water retention is no longer coordinated with sodium retention, the result is hyponatremia. Hyponatremia is more likely to develop in patients with diminished free water clearance at the start of NSAID therapy, e.g. in renal failure, during diuretic treatment, or in congestive heart failure.

The efficacy of a number of *antihypertensive agents* (β-blockers, diuretics, and ACE inhibitors) is blunted by NSAIDs [12]. The duration and extent of this interaction is variable, but no studies have investigated whether the interaction is sustained for more than a few weeks.

96

The incidence of unwanted renal effects of NSAIDs is low (<1% of chroni-
cally-treated patients), [23,24]. Even with study populations ranging from 1,000
to >70,000 patients, elevations of serum creatinine are only found in isolated cases.
Fox et al. [25] identified 1,222 consecutive in-patients receiving NSAIDs (time
and dose, not defined; mean hospital stay, 12 days). A rise in blood urea was found
in 1.1% of these cases, compared to 1.3% in 40,196 controls receiving no NSAIDs.
The authors also followed about 70,000 out-patients on NSAID treatment (chronic
treatment in about 5% of the out-patients). No case of hospital admission for acute
renal disease was reported. Bonney and coworkers [26] published a meta-analysis
of the renal side-effects of two NSAIDs in several double-blind trials. Drug
treatment lasted for 0.5–1 year (oxaprozin, n = 847; aspirin, n = 439; ibuprofen,
n = 182). Prospective measurements identified 3 patients in which serum creatinine increased to over 2 mg/dl, all of whom were on diuretics. Less serious
increases in serum creatinine or blood urea nitrogen were detected in 4–6% of the
patients, without any differences between the three drugs. Placebo treatment was
not included in these protocols.

Nephrology departments in tertiary referral centers may see 2 to 4 cases of
NSAID-related acute renal failure per year. Nevertheless, many patients on
NSAID therapy may experience substantial reductions in GFR without their
serum creatinine titers exceeding the normal range. This is especially true in
patients with reduced skeletal muscle mass, which frequently occurs in rheumatologic diseases. A fall in GFR from, e.g., 110 ml/min to 70 ml/min may be
reflected by an increase in serum creatinine from 0.9 to 1.2 mg/dl. Such an increase
is likely to go unnoticed in most clinical settings. Therefore, the incidence of such
NSAID-induced mild renal impairment is unknown, as are its long-term consequences for the kidney.

Possible differences in renal effects of NSAIDs

Higgs et al. [4] established a rank order of potency of NSAIDs with respect to
inhibition of prostanoid synthesis *in vivo*. This rank order paralleled the relative
ability of these drugs to reduce inflammatory edema formation *in vivo*, but there
is no relationship between the rank potency of prostanoid inhibitors and the
probability of adverse effects. For example, the high incidence of gastrointestinal
complications due to NSAIDs can be markedly reduced by using non-ulcerogenic
sodium salicylate, which reduces prostanoid formation in inflamed tissues without
affecting gastric mucosal cyclooxygenase [27]. Some NSAIDs have also been
found to exhibit differential effects on myelomonocytic migration [27a,28] and
on lipoxygenase activity [29–31]. These results imply that the development of
well-tolerated NSAIDs is feasible.

"Renal Sparing"

The possibility of "renal-sparing" by NSAIDs, mainly aspirin and sulindac, has been well discussed in the medical literature.

Low-dose aspirin (<100 mg/d/70 kg) inhibits platelet cyclooxygenase while the renal enzymic activity remains normal due to continued synthesis of cyclooxygenase. Higher doses of aspirin (e.g. 1 g i.v.) will affect the kidney as discussed above for other NSAIDs.

There has been a widespread interest in sulindac since the demonstration of a renal sparing effect for this drug, which has an anti-inflammatory potency comparable to that of indomethacin [32]. An indoleacetic acid derivative, sulindac is a prodrug which is rapidly converted to a sulfide metabolite that is the likely pharmacophore [33]. Ciabattoni et al. (1980) showed that, in contrast to indomethacin, renal prostanoid synthesis and furosemide-induced renin release are not altered by sulindac either under basal conditions in female volunteers or under conditions of chronically-enhanced prostanoid synthesis such as Bartter's syndrome. Ciabattoni et al. [34] later showed a consistent reduction in GFR in patients with chronic glomerular diseases treated with ibuprofen (1.2 g/d) but not with sulindac (0.4 g/d). Serum TXB_2 was similarly reduced by both drugs, while urinary excretion of 6-keto-$PGF_{1\alpha}$ and PGE_2 was blunted by ibuprofen only.

A recent study on rheumatologic patients with mild azotemia of unknown origin reported a significant increase in serum creatinine titers in 3 of 12 patients on ibuprofen. This AR did not occur when the same patients were challenged with sulindac and piroxicam (11-day treatment). Urinary PG excretion was similarly reduced by all 3 drugs [35]. Analysis of plasma pharmacokinetics indicated that sulindac and piroxicam continued to accumulate even after 11 days of treatment. In other high-risk groups the potential of sulindac to inhibit renal PG production and to impair GFR has been demonstrated. Quintero et al. [36] reported that a 3-day course of sulindac (0.4 g/d) resulted in a ca. 40% reduction in GFR in five patients with decompensated cirrhosis of the liver; plasma concentrations of sulindac in these patients were about four times higher than control values. However, Laffi et al. [37] were unable to demonstrate a reduction in GFR in patients with decompensated cirrhosis of the liver during 5 days of treatment with 0.4 g/d of sulindac despite a reduction in PGE_2 but not in 6-keto-$PGF_{1\alpha}$.

A large number of studies comparing the renal effects of sulindac with other NSAIDs in various patient populations (congestive heart failure, nephrotic syndrome, cirrhosis, treated hypertensives, etc.) have been published. Most of the data (Tab. 4) indicate that sulindac affects renal function less than other NSAIDs, at least the doses and over the timespans investigated.

The aforecited clinical data and the correlation between effective inhibition of

98

Table 4. Evidence in favor and against a renal-sparing effect of sulindac.

Author	Paradigm	Reference drug
In Favor		
Ebel [41] Koopmans [42] Lewis [43] Mills [44] Puddey [45] Salvetti [59,60] Steiness [46] Wong [47]	Interference with antihypertensive drugs	Naproxen Piroxicam Indomethacin
Vriesendorp [48]	Nephrotic syndrome	Indomethacin
Ciabattoni [34]	Glomerulonephritis	Ibuprofen
Kamper [49]	Glomerulonephritis	Naproxen
Whelton [35]	Chronic renal failure	Ibuprofen
Laffi [37]	Liver cirrhosis	Ibuprofen
Daskalopoulos [50]	Liver cirrhosis	Indomethacin
Guarner [51]	Liver cirrhosis	None
Eriksson [52]	Heart failure	Naproxen
Beermann [53]	Heart failure	Naproxen
Sedor [54]	Natriuresis	Indomethacin
Ciabattoni [61]	Volunteers	None
Mistry [55]	5/6 nephrectomy	Indomethacin
Against		
Blackshear [56]	1 patient	None
Corwin [57]	1 patient	None
Quintero [36]	Liver cirrhosis	None
Zambraski [58]	Liver disease in dogs	None
Brater [62]	Liver disease in dogs	Naproxen
Henrich [63]	Hemorrhage in dogs	Indomethacin
Brater [64]	Volunteers	Indomethacin

platelet cyclooxygenase and preservation of renal cyclooxygenase activity after sulindac administration [32] suggests several plausible explanations:

1. the existence of a lower sensitivity of the renal enzyme to the inhibitory action of the sulfide metabolite was hypothesized. In fact, there is strong evidence that different aspirin-like drugs exhibit selective inhibition of cyclooxygenases from different tissues [38];
2. a differential distribution and formation of the sulfide metabolite within the kidney is possible. Segregation of the cyclooxygenase in different renal compartments has been proposed [39];
3. there is some evidence for inactivation of the active moiety of sulindac in renal tissue [12].

Conclusions

The incidence of severe renal ARs due to NSAIDs is low and generally restricted to specific risk groups. In contrast, the incidence of mild renal impairment and its long-term consequences are largely unknown, but careful small sample size studies indicate that such ARs may be common. Low-dose aspirin is "renal sparing" and aspirin use is not associated with analgesic nephropathy.

The bulk of evidence indicates that sulindac is less inhibitory towards renal PG formation and renal hemodynamics (GFR, renal blood flow) than other NSAIDs under many but not all circumstances. In particular, GFR may be reduced by sulindac in cases of circulatory compromise. However, in almost all comparative studies (Tab. 4), fixed doses of sulindac and other NSAIDs were used for 1–2 weeks. It is entirely possible that the renal sparing effect would *not* be observed with higher doses of sulindac or with longer duration of treatment. In fact, Roberts et al. [40] reasoned that sulindac is a generally weaker cyclooxygenase inhibitor (not only for the kidney) than other NSAIDs. Dose-response relationships must obviously be established when the relative potencies of several drugs are to be compared. However, this proposal can only rarely be followed in clinical research.

References

1. Vane, J. and Botting, R., Inflammation and the mechanism of action of anti-inflammatory drugs. FASEB J. **1**: 89–96 (1987).
2. Clive, D.M. and Stoff, J.S., Renal syndromes associated with non-steroidal anti-inflammatory drugs. N. Engl. J. Med. **310**: 563–572 (1984).
3. Huskisson, E.C,. Classification of anti-rheumatic drugs. In: Huskisson, E.C. (Ed.), Anti-rheumatic drugs. Praeger Publishers, New York 1983, pp. 1–9.

4. Higgs, G.A., Harvey, E.A., Ferreira, S.H., and Vane, J.R., The effects of antiinflammatory drugs on the production of prostaglandins *in vivo*. Adv. Prostaglandin Thromboxane Leukotriene Res. **1**: 105–110 (1976).

5. Kulmacz, R.J. and Lands, W.E.M., Stoichiometry and kinetics of the interaction of prostaglandin H synthase with anti-inflammatory agents. J. Biol. Chem. **260**: 12572–12578 (1985).

6. Halliwell, B., Hoult, J.R., and Blake, D.R., Oxidants, inflammation, and antiinflammatory drugs. FASEB J. **2**: 2867–2873.

7. Abrahamson, S.B., Chersky, B., Gude, D., et al. Non-steroidal anti-inflammatory drugs exert differential effects on neutrophil function and plasma membrane viscosity. Inflammation **14**: 11–29 (1990).

8. Becker, E.L., Kermode, J.C., Naccache, P.H. et al. Pertussis toxin as a probe of neutrophil activation. Fed. Proc. **45**: 2151–2155 (1986).

9. Abrahamson, S., Cherksky, B., Veiro, D. et al., Nonsteroidal antiinflammatory agents, but not acetaminophen, disrupt signal transduction within the neutrophil plasmalemma. Fed. Proc. **46**: 1389 (1987).

10. Crowell, R.E. and Vaan Epps, D.E., Non-steroidal anti-inflammatory agents inhibt upregulation of CDllb, CDllc, and CD35 in neutrophils stimulated by formyl-methionine-leucine-phenylalanine. Inflammation **14**: 163–171 (1990).

11. Levine, R.A., Nandi, J., and King, R.L., Aspirin potentiates prestimulated acid secretion and mobilizes intracellular calcium in rabbit parietal cells. J. Clin. Invest. **86**: 400–408 (1990).

12. Patrono, C. and Dunn, M.J., The clinical significance of inhibition of renal prostaglandin synthesis. Kidney Int. **32**: 1–12 (1987).

13. Needleman, P., Turk, J., Jakschik, A. et al., Arachidonic acid metabolism. Annu. Rev. Biochem. **55**: 69–102 (1986).

14. Gardner, D.G. and Schultz, H.D., Prostaglandins regulate the synthesis and secretion of the atrial natriuretic peptide. J. Clin. Invest. **86**: 52–59 (1990).

15. Marshall, P.J., Kulmacz, R.J., and Lands, W.E.M., Constraints on prostaglandin biosynthesis in tissues. J. Biol. Chem. **262**: 3510–5317 (1987).

15a. Oates, J.A., FitzGerald, G.A., Branch, R.A. et al., Clinical implications of prostaglandin and thromboxane A_2 formation. N. Engl. J. Med. **319**: 761–767 (1988).

16. Oberle, G.P. and Stahl, R.A.K., Akute Nebenwirkungen nicht-steroidaler Antiphlogistika auf die Nieren. Dtsch. Med. Wochenschr. **115**: 309–314 (1990).

17. Koeppen-Hagemann, I., Binkele-Uihlein, U., Waldherr, R., Andrassy, K. et al., Akute granulomatöse interstitielle Nephritis mit Iritis. Dtsch. Med. Wochenschr. **112**: 259–261 (1987).

18. Sandler, D.P., Smith, J.C., Weinberg, C.R. et al., Analgesic use and chronic renal disease. N. Engl. J. Med. **320**: 1238–1242 (1989).

19. Bennett, W.M. and DeBroe, M.E., Analgesic nephropathy–a preventable renal disease. N. Engl. J. Med. **320**: 1269–1271 (1989).

20. Adams, D.H., Howie, A.J., Michael, J., McConkey, B. et al. Nonsteroidal antiinflammatory drugs and renal failure. Lancet **1**:57–60 (1986).

20a. Tan, S.Y., Shapiro. R., and Ksih, M.A., Reversible acute renal failure induced by indomethacin. J. Am. Med. Assn. **241**: 2732–2733 (1979).

21. Feldman, D., Loose, D.S., and Tan, S.Y., Nonsteroidal antiinflammatory drugs cause sodium and water retention in the rat. Am. J. Physiol. **234**: F490–F4966 (1978).

21a Brater, D.C., Effects of indomethacin on salt and water homeostasis. Clin. Pharmacol. Ther. **25**: 322–330 (1979).

22. Patak, R.V., Mookerjee, B.K., Bentzel, C.J., Hysert, P.E. et al., Antagonism of the effects of furosemide by indomethacin in normal and hypertensive man. Prostaglandins **10**: 649–659 (1975).
23. Murray, M.D. and Brater, D.C., Adverse effects of non-steroidal antiinflammatory drugs on renal function. Ann. Intern. Med. **112**: 559–560 (1990).
24. Kurowski, M., SPALA – Sicherheitsprofil von Antirheumatika bei Langzeitanwendung. Dtsch. Ärzteblatt **87**: 1916–1924 (1990).
25. Fox, D.A. and Jick, H., Non-steroidal anti-inflammatory drugs and renal disease. J. Am. Med. Assn. **251**: 1299–1300 (1984).
26. Bonney, S.L., Northington, R.S., Hedrich, D.A., and Walker, B.R., Renal safety of two analgesics used over the counter: ibuprofen and aspirin. Clin. Pharmacol. Ther. **40**: 373–377 (1986).
27. Whittle, B.J.R., Higgs, G.A., Eakins, K.E. et al., Selective inhibition of prostaglandin production in inflammatory exudates and gastric mucosa. Nature **284**: 271–273 (1980).
27a.Higgs, G.A., Eakins, K.E., Mugridge, K.G., et al., The effects of nonsteroid antiinflammtory drugs on leukocyte migration in carrageen-induced inflammation. Eur. J. Pharmacol. **66**: 81–86 (1980).
28. Adams, S.S., Burrows, C.A., Sheldon, N., and Yates, D.B., Inhibition of prostaglandin synthesis and leukocyte migration by flurbiprofen. Curr. Med. Res. **5**: 11–16 (1977).
29. Bragt, P.C. and Bonta, I.L., Indomethacin inhibits the formation of the lipoxygenase product HETE during granulomatous inflammation in the rat. J. Pharm. Pharmacol. **32**: 143–144 (1980).
30. Koshihara, Y., Nagasaki, I., and Murota, S.I., Production of slow reacting substance in rat granulomatous inflammation. Biochem. Pharmacol. **30**: 1781–1784 (1981).
31. Randall, R.W., Eakins, K.E., Higgs, G.A. et al. Inhibition of arachidonic acid cyclo-oxygenase and lipoxygenase activities of leukocytes by indomethacin and compound BW755C. Agents Actions **10**: 553–555 (1980).
32. Rhymer, A.R., Sulindac. In: Huskisson, E.C. (Ed.), Anti-rheumatic drugs. Praeger Publishers, New York 1983, pp. 421–4437.
33. Kwan, K.C., Duggan, D.E., Van Arman, E.G., and Shen, T.Y,. Sulindac: chemistry, pharmacology and pharmacokinetics. Eur. J. Rheumatol. Inflamm. **1**: 9–11 (1978).
33a.Ciabattoni, G., Pugliese, F., Cinotti, G. A., and Patrono, C., Renal effects of antiinflammatory drugs. Eur. J. Rheumatol., **3**: 210–212 (1980).
34. Ciabattoni, G., Cinotti, G.A., Pierucci, A. et al., Effects of sulindac and ibuprofen in patients with chronic glomerular disease. Evidence for the dependence of renal function on prostacyclin. N. Engl. J. Med. **310**: 279–283 (1984).
35. Whelton, A., Stout, R.L., Spilman, P.S., and Klassen, D.K., Renal effects of ibuprofen, piroxicam, and sulindac in patients with asymtomatic renal failure. A prospective, randomized, crossover comparison. Ann. Intern. Med. **112**: 568–576 (1990).
36. Quintero, E., Gines, P., Arroyo, V., Rimola, A. et al., Sulindac reduces the urinary excretion of prostaglandins and impairs renal function in cirrhosis with ascites. Nephron **42**: 298–303 (1986).
37. Laffi. G., La Villa, G., Pinzani, M., Ciabattoni, G. et al., Altered and renal and platelet arachidonic acid metabolism in cirrhosis. Gastroenterology **90**: 274–282 (1986).
38. Flower, R.J. and Blackwell, G.J., Antiinflammatory steroids induce biosynthesis of a phospholipase inhibitor which prevents prostaglandin generation. Nature **278**: 456–459 (1979).
39. McGiff, J.C. and Wong, P.Y., Compartmentalization of prostaglandins and prostacyclin within the kidney: implications for renal function. Fed. Proc. **38**: 89–93 (1979).

40. Roberts, D.G., Gerber, J.G., Barnes, J.S., Zebre, G.O. et al., Sulindac is not renal-sparing in man. Clin. Pharmacol. Ther. **38**: 258–265 (1985).

41. Ebel, D.L., Rhymer, A.R., and Stahl, E., Effects of sulindac, piroxicam and placebo on the hypotensive effect of propranolol in patients with mild to moderate essential hypertension. Adv. Ther. **2**: 131–142 (1985).

42. Koopmans, P.P., Thien, T.H., and Gribnau, F.W.J., Influence of nonsteroidal antiinflammtory drugs on diuretic treatment of mild to moderate essential hypertension. Br. Med. J. **289**: 1492–1494 (1984).

43. Lewis, R.V., Toner, J.M., Jackson, P.R., and Ramsay L.E., Effects of indomethacin and sulindac on blood pressure of hypertensive patients. Br. Med. J. **292**: 934–935 (1986).

44. Mills, E.H., Whitworth, J.A., Andrews, J., and Kincaid-Smith, P., Nonsteroidal antiinflammatory drugs and blood pressure. Aust. N. Z. J. Med. **12**: 478–482 (1982).

45. Puddey, I.B., Beilin, L.J., Vandongen, R., Banks, R. et al., Differential effects of sulindac and indomethacin on blood pressure in treated essential hypertensive subjects. Clin. Sci. **69**: 327–336 (1985).

46. Steiness, E. and Waldorff, S., Different interaction of indomethacin and sulindac with thiazides in hypertension. Br. Med. J. **285**: 1702–1703 (1982).

47. Wong, D.G., Spence, J.D., Lamki, L., Freeman, D. et al., Effect of nonsteroidal antiinflammatory drugs on control of hypertension by beta blockers and diuretics. Lancet **i**:997–1001 (1986).

48. Vriesendorp, R., De Zweeuw, D., De Jong, P.E., Donker, A.J.M. et al., Reduction of urinary protein and prostaglandin E_2 excretion in the nephrotic syndrome by nonsteroidal antiinflammatory drugs. Clin. Nephrol. **25**: 105–110 (1986).

49. Kamper, A.L., Strandgaard, S., Christensen, P., and Svendsen, U.G., Effects of sulindac and naproxen in patients with chronic glomerular disease. Scand. J. Rheumatol. **62** (Suppl): 26 (1986).

50. Daskalopoulos, G., Kronborg, I., Katkov, W., Gonzales, M. et al., Sulindac and indomethacin suppress the diuretic action of furosemide in patients with cirrhosis and ascites: evidence that sulindac affects renal prostaglandins. Am. J. Kidney Dis. **6**: 217–221 (1985).

51. Guarner, C., Enriquez, F Guarner et al. Effects of sulindac on prostaglandin excretion and renal function in cirrhotic patients with ascites. Scand. J. Gastroenterol. **21**: 231–234 (1986).

52. Eriksson, L.O., Beermann, B., and Kallner, M., Renal function and tubular transport effects of sulindac and naproxen in chronic heart failure. Clin. Pharmacol. Ther. **42**: 646–654 (1987).

53. Beermann, B., Eriksson, L.O., and Kallner, M., A double blind comparison of naproxen and sulindac in female patients with heart failure. Scand. J. Rheumatol. **62**: 32 (1986).

54. Sedor, J.R., Davidson, E., and Dunn, M.J., Effects of nonsteroidal antiinflammatory drugs in healthy subjects. Am. J. Med. **81**: 58–70 (1986).

55. Mistry, C.D., Lote, C.J., Gokal, R., Currie, W.J. et al., Effects of sulindac on renal function and prostaglandin synthesis in patients with moderate chronic renal insufficiency. Clin. Sci. **70**: 501–505 (1986).

56. Blackshear, J.L., Davidman, M., and Stillmann, T., Identification of risk for renal insufficiency from non-steroidal anti-inflammatory drugs. Arch. Intern. Med. **143**: 1131 (1983).

57. Corwin, H.K. and Bonventre, J.V., Renal insufficiency associated with nonsteroidal antiinflammatory agents. Am. J. Kidney Dis. **4**: 147 (1984).

58. Zambraski, E.J., Chremos, A.N., and Dunn, M.J., Comparison of the effects of sulindac with other cyclooxygenase inhihitors on prostaglandin excretion and renal function in normal and chronic bile duct-ligated dogs and swine. J. Pharmacol. Exp. Ther. **228**: 560–566 (1984).

59. Salvetti, A., Pedrinelli, R., Magagna, A. and Ugenti, P., Differential effects of selective and non-selective prostaglandin synthesis inhibition on the pharmacology responses to captopril in patients with essential hypertension. Clin. Sci. **63**: 261S–263S (1982).
60. Salvetti, A., Pedrinelli, R., Alberici, P., Magagna, A. et al. The influence of indomethacin and sulindac on some pharmacological actions of atenolol in hypertensive patients. Br. J. Clin. Pharmacol. **17**: 108S–111S (1984).
61. Ciabattoni, G., Boss, A.H., Patrignani, P., Catella, F. et al. Dose-dependent inhibition of extra-renal prostacyclin production by sulindac in man. Clin. Pharmacol. Ther. **41**: 380–383 (1987).
62. Brater, D.C., Anderson, S.A., and Brown-Cartwright, D., Reversible acute decrease in renal function by NSAIDs in cirrhosis. Am. J. Med. Sci. **294**: 168–174 (1987).
63. Henrich, W.L., Brater, D.C., and Campell, W.B., Renal hemodynamic effects of therapeutic plasma levels of sulindac sulfide during hemorrhage. Kidney Int. **29**: 484–489 (1986).
64. Brater, D.C., Anderson, S., Baird, B., and Campbell, W.B., Effects of ibuprofen, naproxen, and sulindac on prostaglandins in men. Kidney Int. **27**: 66–73 (1985).